IMPROVING HEALTH IN THE UNITED STATES

The Role of
Health Impact Assessment

Committee on Health Impact Assessment

Board on Environmental Studies and Toxicology

Division on Earth and Life Studies

National Research Council

NATIONAL RESEARCH COUNCIL
OF THE NATIONAL ACADEMIES

THE NATIONAL ACADEMIES PRESS
Washington, D.C.
www.nap.edu

THE NATIONAL ACADEMIES PRESS   500 Fifth Street, NW   Washington, DC 20001

NOTICE: The project that is the subject of this report was approved by the Governing Board of the National Research Council, whose members are drawn from the councils of the National Academy of Sciences, the National Academy of Engineering, and the Institute of Medicine. The members of the committee responsible for the report were chosen for their special competences and with regard for appropriate balance.

This project was supported by contracts between the National Academy of Sciences and Robert Wood Johnson Foundation, Grant No. 66737; The California Endowment, Grant No. 20091397; DHHS/CDC, Contract No. 200-2005-13434; and DHHS/NIH, Contract No. N01-OD-4-2139. Any opinions, findings, conclusions, or recommendations expressed in this publication are those of the authors and do not necessarily reflect the view of the organizations or agencies that provided support for this project.

International Standard Book Number-13: 978-0-309-21883-2
International Standard Book Number-10: 0-309-21883-7
Library of Congress Control Number: 2011939904

Additional copies of this report are available from

The National Academies Press
500 Fifth Street, NW
Box 285
Washington, DC 20055

800-624-6242
202-334-3313 (in the Washington metropolitan area)
http://www.nap.edu

Copyright 2011 by the National Academy of Sciences. All rights reserved.

Printed in the United States of America.

THE NATIONAL ACADEMIES
Advisers to the Nation on Science, Engineering, and Medicine

The **National Academy of Sciences** is a private, nonprofit, self-perpetuating society of distinguished scholars engaged in scientific and engineering research, dedicated to the furtherance of science and technology and to their use for the general welfare. Upon the authority of the charter granted to it by the Congress in 1863, the Academy has a mandate that requires it to advise the federal government on scientific and technical matters. Dr. Ralph J. Cicerone is president of the National Academy of Sciences.

The **National Academy of Engineering** was established in 1964, under the charter of the National Academy of Sciences, as a parallel organization of outstanding engineers. It is autonomous in its administration and in the selection of its members, sharing with the National Academy of Sciences the responsibility for advising the federal government. The National Academy of Engineering also sponsors engineering programs aimed at meeting national needs, encourages education and research, and recognizes the superior achievements of engineers. Dr. Charles M. Vest is president of the National Academy of Engineering.

The **Institute of Medicine** was established in 1970 by the National Academy of Sciences to secure the services of eminent members of appropriate professions in the examination of policy matters pertaining to the health of the public. The Institute acts under the responsibility given to the National Academy of Sciences by its congressional charter to be an adviser to the federal government and, upon its own initiative, to identify issues of medical care, research, and education. Dr. Harvey V. Fineberg is president of the Institute of Medicine.

The **National Research Council** was organized by the National Academy of Sciences in 1916 to associate the broad community of science and technology with the Academy's purposes of furthering knowledge and advising the federal government. Functioning in accordance with general policies determined by the Academy, the Council has become the principal operating agency of both the National Academy of Sciences and the National Academy of Engineering in providing services to the government, the public, and the scientific and engineering communities. The Council is administered jointly by both Academies and the Institute of Medicine. Dr. Ralph J. Cicerone and Dr. Charles M. Vest are chair and vice chair, respectively, of the National Research Council.

www.national-academies.org

COMMITTEE ON HEALTH IMPACT ASSESSMENT

Members

RICHARD J. JACKSON (*Chair*), University of California, Los Angeles
DINAH BEAR, Attorney at Law, Washington, DC
RAJIV BHATIA, San Francisco Department of Public Health; University of California, San Francisco
SCOTT B. CANTOR, The University of Texas MD Anderson Cancer Center, Houston
BEN CAVE, Ben Cave Associates, Ltd., Leeds, United Kingdom
ANA V. DIEZ ROUX, University of Michigan, Ann Arbor
CARLOS DORA, World Health Organization, Geneva, Switzerland
JONATHAN E. FIELDING, Los Angeles County Department of Public Health, Los Angeles, CA
JOSHUA S. GRAFF ZIVIN, University of California, San Diego
JONATHAN I. LEVY, Boston University School of Public Health, Boston, MA
JULIA B. QUINT, California Department of Public Health (retired), Berkeley
SAMINA RAJA, University at Buffalo, State University of New York, Buffalo
AMY JO SCHULZ, University of Michigan, Ann Arbor
AARON A. WERNHAM, Pew Charitable Trusts, Washington, DC

Staff

ELLEN K. MANTUS, Project Director
HEIDI MURRAY-SMITH, Program Officer
KERI SCHAFFER, Research Associate
NORMAN GROSSBLATT, Senior Editor
MIRSADA KARALIC-LONCAREVIC, Manager, Technical Information Center
RADIAH ROSE, Manager, Editorial Projects
PANOLA GOLSON, Program Associate

Sponsors

ROBERT WOOD JOHNSON FOUNDATION
CALIFORNIA ENDOWMENT
NATIONAL INSTITUTE OF ENVIRONMENTAL HEALTH SCIENCES
U.S. CENTERS FOR DISEASE CONTROL AND PREVENTION

BOARD ON ENVIRONMENTAL STUDIES AND TOXICOLOGY[1]

Members

ROGENE F. HENDERSON (*Chair*), Lovelace Respiratory Research Institute, Albuquerque, NM
PRAVEEN AMAR, Clean Air Task Force, Boston, MA
TINA BAHADORI, American Chemistry Council, Washington, DC
MICHAEL J. BRADLEY, M.J. Bradley & Associates, Concord, MA
JONATHAN Z. CANNON, University of Virginia, Charlottesville
GAIL CHARNLEY, HealthRisk Strategies, Washington, DC
FRANK W. DAVIS, University of California, Santa Barbara
RICHARD A. DENISON, Environmental Defense Fund, Washington, DC
CHARLES T. DRISCOLL, JR., Syracuse University, New York
H. CHRISTOPHER FREY, North Carolina State University, Raleigh
RICHARD M. GOLD, Holland & Knight, LLP, Washington, DC
LYNN R. GOLDMAN, George Washington University, Washington, DC
LINDA E. GREER, Natural Resources Defense Council, Washington, DC
WILLIAM E. HALPERIN, University of Medicine and Dentistry of New Jersey, Newark
PHILIP K. HOPKE, Clarkson University, Potsdam, NY
HOWARD HU, University of Michigan, Ann Arbor
SAMUEL KACEW, University of Ottawa, Ontario
ROGER E. KASPERSON, Clark University, Worcester, MA
THOMAS E. MCKONE, University of California, Berkeley
TERRY L. MEDLEY, E.I. du Pont de Nemours & Company, Wilmington, DE
JANA MILFORD, University of Colorado at Boulder, Boulder
FRANK O'DONNELL, Clean Air Watch, Washington, DC
RICHARD L. POIROT, Vermont Department of Environmental Conservation, Waterbury
KATHRYN G. SESSIONS, Health and Environmental Funders Network, Bethesda, MD
JOYCE S. TSUJI, Exponent Environmental Group, Bellevue, WA

Senior Staff

JAMES J. REISA, Director
DAVID J. POLICANSKY, Scholar
RAYMOND A. WASSEL, Senior Program Officer for Environmental Studies
SUSAN N.J. MARTEL, Senior Program Officer for Toxicology
ELLEN K. MANTUS, Senior Program Officer for Risk Analysis
EILEEN N. ABT, Senior Program Officer
RUTH E. CROSSGROVE, Senior Editor
MIRSADA KARALIC-LONCAREVIC, Manager, Technical Information Center
RADIAH ROSE, Manager, Editorial Projects

---

[1] This study was planned, overseen, and supported by the Board on Environmental Studies and Toxicology.

OTHER REPORTS OF THE
BOARD ON ENVIRONMENTAL STUDIES AND TOXICOLOGY

A Risk-Characterization Framework for Decision-Making at the Food and Drug
    Administration (2011)
Review of the Environmental Protection Agency's Draft IRIS Assessment of
    Formaldehyde (2011)
Toxicity-Pathway-Based Risk Assessment: Preparing for Paradigm Change (2010)
The Use of Title 42 Authority at the U.S. Environmental Protection Agency (2010)
Review of the Environmental Protection Agency's Draft IRIS Assessment of
    Tetrachloroethylene (2010)
Hidden Costs of Energy: Unpriced Consequences of Energy Production and Use (2009)
Contaminated Water Supplies at Camp Lejeune—Assessing Potential Health
    Effects (2009)
Review of the Federal Strategy for Nanotechnology-Related Environmental, Health,
    and Safety Research (2009)
Science and Decisions: Advancing Risk Assessment (2009)
Phthalates and Cumulative Risk Assessment: The Tasks Ahead (2008)
Estimating Mortality Risk Reduction and Economic Benefits from Controlling Ozone
    Air Pollution (2008)
Respiratory Diseases Research at NIOSH (2008)
Evaluating Research Efficiency in the U.S. Environmental Protection Agency (2008)
Hydrology, Ecology, and Fishes of the Klamath River Basin (2008)
Applications of Toxicogenomic Technologies to Predictive Toxicology and Risk
    Assessment (2007)
Models in Environmental Regulatory Decision Making (2007)
Toxicity Testing in the Twenty-first Century: A Vision and a Strategy (2007)
Sediment Dredging at Superfund Megasites: Assessing the Effectiveness (2007)
Environmental Impacts of Wind-Energy Projects (2007)
Scientific Review of the Proposed Risk Assessment Bulletin from the Office of
    Management and Budget (2007)
Assessing the Human Health Risks of Trichloroethylene: Key Scientific Issues (2006)
New Source Review for Stationary Sources of Air Pollution (2006)
Human Biomonitoring for Environmental Chemicals (2006)
Health Risks from Dioxin and Related Compounds: Evaluation of the EPA
    Reassessment (2006)
Fluoride in Drinking Water: A Scientific Review of EPA's Standards (2006)
State and Federal Standards for Mobile-Source Emissions (2006)
Superfund and Mining Megasites—Lessons from the Coeur d'Alene River Basin (2005)
Health Implications of Perchlorate Ingestion (2005)
Air Quality Management in the United States (2004)
Endangered and Threatened Species of the Platte River (2004)
Atlantic Salmon in Maine (2004)
Endangered and Threatened Fishes in the Klamath River Basin (2004)
Cumulative Environmental Effects of Alaska North Slope Oil and Gas
    Development (2003)
Estimating the Public Health Benefits of Proposed Air Pollution Regulations (2002)
Biosolids Applied to Land: Advancing Standards and Practices (2002)

The Airliner Cabin Environment and Health of Passengers and Crew (2002)
Arsenic in Drinking Water: 2001 Update (2001)
Evaluating Vehicle Emissions Inspection and Maintenance Programs (2001)
Compensating for Wetland Losses Under the Clean Water Act (2001)
A Risk-Management Strategy for PCB-Contaminated Sediments (2001)
Acute Exposure Guideline Levels for Selected Airborne Chemicals (ten
    volumes, 2000-2011)
Toxicological Effects of Methylmercury (2000)
Strengthening Science at the U.S. Environmental Protection Agency (2000)
Scientific Frontiers in Developmental Toxicology and Risk Assessment (2000)
Ecological Indicators for the Nation (2000)
Waste Incineration and Public Health (2000)
Hormonally Active Agents in the Environment (1999)
Research Priorities for Airborne Particulate Matter (four volumes, 1998-2004)
The National Research Council's Committee on Toxicology: The First 50 Years (1997)
Carcinogens and Anticarcinogens in the Human Diet (1996)
Upstream: Salmon and Society in the Pacific Northwest (1996)
Science and the Endangered Species Act (1995)
Wetlands: Characteristics and Boundaries (1995)
Biologic Markers (five volumes, 1989-1995)
Science and Judgment in Risk Assessment (1994)
Pesticides in the Diets of Infants and Children (1993)
Dolphins and the Tuna Industry (1992)
Science and the National Parks (1992)
Human Exposure Assessment for Airborne Pollutants (1991)
Rethinking the Ozone Problem in Urban and Regional Air Pollution (1991)
Decline of the Sea Turtles (1990)

Copies of these reports may be ordered from the National Academies Press
(800) 624-6242 or (202) 334-3313
www.nap.edu

Preface

A growing body of evidence indicates that many factors outside the traditional health field affect public health. The idea that our health is determined only by our own behavior, choices, and genetics is no longer tenable. Many now recognize that substantial improvements in public health will occur only by ensuring that health considerations are factored into projects, programs, plans, and policies in non-health-related sectors, such as transportation, housing, agriculture, and education.

Health impact assessment (HIA) is a tool that can help decision-makers identify the public-health consequences of proposals that potentially affect health. Because of the potential that HIA offers to improve public health, the Robert Wood Johnson Foundation, the National Institute of Environmental Health Sciences, the California Endowment, and the U.S. Centers for Disease Control and Prevention asked the National Research Council to develop a framework, terminology, and guidance for conducting HIA.

In this report, the Committee on Health Impact Assessment discusses the need for health-informed decision-making and policies and reviews the current practice of HIA. The committee provides a definition, framework, and criteria for HIA; discusses issues in and challenges to the development and practice of HIA; and closes with a discussion on structures and policies for promoting HIA. The committee notes that the framework provided in this report is not a reinvention of the field but a synthesis of guidance provided in other documents and publications. Thus, the reader will find many similarities between the committee's descriptions and characterizations and those of other guides.

The present report has been reviewed in draft form by persons chosen for their diverse perspectives and technical expertise in accordance with procedures approved by the National Research Council Report Review Committee. The purpose of the independent review is to provide candid and critical comments that will assist the institution in making its published report as sound as possible and to ensure that the report meets institutional standards of objectivity, evidence, and responsiveness to the study charge. The review comments and draft manuscript remain confidential to protect the integrity of the deliberative process. We thank the following for their review of this report: Jason Corburn, University of California, Berkeley; William H. Dow, University of California, Berkeley; Jonathan C. Heller, Human Impact Partners; Murray Lee, Habitat Health Impact Consulting; Jonathan Levine, University of Michigan; Linda A. McCauley, Emory University; David O. Meltzer, University of Chicago; Keshia M. Pollack, Johns Hopkins Bloomberg School of Public Health; Lindsay Rosenfeld, Northeastern University; Alex Scott-Samuel, University of Liverpool; Nicholas C. Yost, SNR Denton; Lauren Zeise, California Environmental Protection Agency.

Although the reviewers listed above have provided many constructive comments and suggestions, they were not asked to endorse the conclusions or recommendations, nor did they see the final draft of the report before its release. The review of the report was overseen by the review coordinator, Joseph V. Rodricks, Environ, and the review monitor, Gilbert S. Omenn, University of Michigan Medical School. Appointed by the National Research Council, they were responsible for making certain that an independent examination of the report was carried out in accordance with institutional procedures and that all review comments were carefully considered. Responsibility for the final content of the report rests entirely with the committee and the institution.

The committee gratefully acknowledges the following for their presentations: Marice Ashe, Public Health Law and Policy; John Balbus, National Institute of Environmental Health Sciences; Ronald Bass, ICF International; Larry Cohen, Prevention Institute; Andrew Dannenberg, U.S. Centers for Disease Control and Prevention; Paul Farmer, American Planning Association; Ed Fogels, Alaska Department of Natural Resources; Robert Gould, Partnership for Prevention; Ralph Keeney, Duke University; Jenelle Krishnamoorthy, U.S. Senate Committee on Health, Education, Labor, and Pensions; Angelo Logan, East Yard Communities for Environmental Justice; April Marchese, U.S. Department of Transportation; John Norquist, Congress for the New Urbanism; Linda Rudolph, California Department of Public Health; Pamela Russo, Robert Wood Johnson Foundation; and Terry Williams, Tulalip Natural Resources Treaty Rights Office.

The committee is also grateful for the assistance of the National Research Council staff in preparing this report. Staff members who contributed to the effort are Ellen Mantus, project director; Heidi Murray-Smith, program officer; Keri Schaffer, research associate; James Reisa, director of the Board on Environmental Studies and Toxicology; Norman Grossblatt, senior editor; Mirsada Karalic-Loncarevic, manager, Technical Information Center; Radiah Rose, manager, editorial projects; and Panola Golson, program associate.

I would especially like to thank the members of the committee for their efforts throughout the development of this report.

Richard J. Jackson, *Chair*
Committee on Health Impact Assessment

Contents

SUMMARY 3

1 INTRODUCTION 14
   Health Impact Assessment, 14
   The Committee's Task and Approach, 18
   Organization of Report, 19
   References, 20

2 WHY WE NEED HEALTH-INFORMED POLICIES
   AND DECISION-MAKING 23
   Knowledge of Root Causes of Health Consequences, 25
   Why Assess the Health Consequences of Policies,
      Programs, Projects, and Planning Decisions?, 27
   Why Assessments Are Not Being Conducted, 30
   What are the Options for Assessment?, 31
   Other Benefits of Systematic Assessment of Health Impacts, 33
   Conclusions, 34
   References, 35

3 ELEMENTS OF A HEALTH IMPACT ASSESSMENT 43
   Categories of Health Impact Assessment, 44
   Definition of Health Impact Assessment, 45
   Who Conducts Health Impact Assessments?, 46
   Process for Health Impact Assessment, 47
   Summary: What Criteria Define a Health Impact Assessment?, 82
   References, 83

4 CURRENT ISSUES AND CHALLENGES IN THE
   DEVELOPMENT AND PRACTICE OF HEALTH
   IMPACT ASSESSMENT 90
   Defining Health for Health Impact Assessment, 90

Are All Decisions Potential Candidates for Health Impact Assessment?, 92
Balancing the Need to Provide Timely, Valid Information with the Realities of Varied Data Quality, 95
Benefits and Challenges of Quantitative Estimation, 99
Characterizing Multiple Health Effects, 101
Assigning Monetary Values to Health Consequences, 102
Valuing and Enabling Stakeholder Participation, 103
The Benefits of a Peer-Review Process for Health Impact Assessment, 106
Minimizing Conflicts of Interest of Sponsors and Practitioners of Health Impact Assessment, 107
Managing Expectations: Information May Not Change Decisions, 108
Advancing Requirements for Health Analysis in Environmental Impact Assessment, 110
Conclusions, 112
References, 114

5   STRUCTURES AND POLICIES FOR PROMOTING HEALTH IMPACT ASSESSMENT 119
Structure and Policies to Support Health Impact Assessment, 120
Promotion of Education and Training in and Societal Awareness of Health Impact Assessment, 124
Increase in Research and Scholarship in Health Impact Assessment, 126
Development of Resources to Support Health Impact Assessment, 128
References, 128

APPENDIXES

APPENDIX A: EXPERIENCES WITH HEALTH IMPACT ASSESSMENT 130

APPENDIX B: BIOGRAPHIC INFORMATION ON THE COMMITTEE ON HEALTH IMPACT ASSESSMENT 178

APPENDIX C: STATEMENT OF TASK OF THE COMMITTEE ON HEALTH IMPACT ASSESSMENT 184

APPENDIX D: GLOSSARY 185

APPENDIX E: SUMMARY OF HEALTH IMPACT ASSESSMENT GUIDES 196

APPENDIX F: ANALYSIS OF HEALTH EFFECTS UNDER THE NATIONAL ENVIRONMENTAL POLICY ACT 204

BOXES, FIGURES, AND TABLES

BOXES

3-1 Screening: HIA of a Residential Housing Program, 50
3-2 Scoping: Atlanta BeltLine HIA, 56
3-3 Assessment: Northeast National Petroleum Reserve-Alaska, 65
3-4 Examples of Health and Behavioral Effects That Have Been Addressed Quantitatively in HIA, 67
3-5 HIA Recommendations, 69
3-6 Reporting: Legislation on Paid Sick Days, 74
A-1 European Union Members and When They Joined, 137

FIGURES

S-1 Framework for HIA, illustrating steps and outputs, 7
3-1 Example of a logic framework that maps out the possible causal pathways by which health effects might occur, 54
A-1 Number of requests for consultation received by the Québec Ministry of Health and Social Services, 2003-2008, 132

TABLES

1-1 Selected Definitions of Health Impact Assessment, 16
2-1 The Costs of Transportation-Related Health Outcomes, 2008, 29
3-1 Example of a Table Used for Systematic Scoping, 55
3-2 Example of a Matrix to Analyze Health Effects, 63
3-3 Example of a Table for Rating Importance of Health Effects, 64
4-1 Health Impact Assessment by Sector, 93
E-1 A Review of Health Impact Assessment Guides, 197
E-2 Health Impact Assessment Guides for Policies or Plans, 200

IMPROVING HEALTH IN THE UNITED STATES

The Role of
Health Impact Assessment

Summary

Many Americans believe that the United States has one of the best health-care systems in the world and that consequently Americans enjoy better health than most of the world's populations. The data, however, do not support that belief. In fact, the United States is ranked 32nd in the world in life expectancy even though it is ranked third in total expenditures on health care as a percentage of gross domestic product (GDP). Clearly, good health is determined by more than money spent on the health-care system. In fact, a growing body of research indicates that living conditions—including such factors as housing quality, exposure to pollution, and access to healthy and affordable foods and safe places to exercise—have a greater effect on health. That research highlights the importance of considering health in developing policies, programs, plans, and projects, including ones that may not appear at first to have an obvious relationship to health.

Health impact assessment (HIA) has arisen as an especially promising way to factor health considerations into the decision-making process. It has been defined in various ways but essentially is a structured process that uses scientific data, professional expertise, and stakeholder input to identify and evaluate public-health consequences of proposals and suggests actions that could be taken to minimize adverse health impacts and optimize beneficial ones. HIA has been used throughout the world to evaluate the potential health consequences of a wide array of proposals that span many sectors and levels of government. International organizations, such as the World Health Organization and multilateral development banks, have also contributed to the development and evolution of HIA, and countries and organizations have both developed their own guidance on conducting HIA.

Although HIA has not been used widely by decision-makers in the United States, its use has steadily increased over the last 10 years. Local, state, and tribal health departments have conducted HIAs to inform decision-making in other agencies; community-based organizations have conducted HIAs with input from public-health experts to inform officials who are deliberating on legislative or administrative proposals; planning and transportation departments have conducted HIAs to inform their own decisions; and private consultants have conducted HIAs for industry to determine the potential health consequences of various projects. Given the potential health benefits of HIA, the Robert Wood Johnson Foundation, the National Institute of Environmental Health Sciences, the California Endowment, and the Centers for Disease Control and Prevention asked the National Research Council (NRC) to develop a framework, terminology, and guidance for conducting HIA of proposed policies, programs, and projects at the federal, state, tribal, and local levels, including the private sector. As a result of that request, NRC convened the Committee on Health Impact Assessment, which prepared this report.

THE NEED FOR HEALTH-INFORMED DECISION-MAKING

The U.S. population clearly has not reached its full health potential despite major medical advances and large expenditures on health care. Almost 50% of adults suffer from at least one chronic illness, and obesity, which contributes to many health conditions, has grown to epidemic proportions in children and adults. Poor health has implications not only for the quality and duration of life but for the economy. Health-care spending accounted for 7% of U.S. GDP in 1970, accounted for 16% of GDP in 2008, and is projected to account for almost 20% by 2019. Poor health also results in reduced participation in and productivity of the labor force. Thus, the consequences of chronic illness are huge in suffering *and* monetary and business costs.

Many scientists, policy-makers, and others recognize that health is determined by multiple factors, including factors that shape the conditions in which people are born, grow, live, work, and age. Policies and programs that have historically not been recognized as related to health are now known or thought to have important health consequences. For example, public health has been linked to housing policies that determine the quality and location of housing developments, to transportation policies that affect the availability of public transportation, to urban planning policies that determine land use and street connectivity, to agricultural policies that influence the availability of various types of food, and to economic-development policies that affect the location of businesses and industry. The recognition that health is shaped by a broad array of factors emphasizes the importance of understanding the possible health consequences of decision-making. In fact, it can be argued that major improvements in public health cannot be achieved without considering the root causes of ill health. Indeed, it has been argued that major health problems, such as the obesity epidemic and its associated health and monetary costs, are essentially unintended consequences of various social and policy factors related, for example, to the mass production and distribution of energy-dense foods and the engineering of physical activity out of daily life through changes in how transportation is organized and how neighborhoods are designed and built.

Accordingly, systematic assessment of the health consequences of policies, programs, plans, and projects is critically important for protecting and promoting public health; as indicated, lack of assessment can have many unexpected adverse health (and economic) consequences. One striking example is development of the transportation infrastructure in the United States. In 1956, Congress passed the Interstate Highway Act, which resulted in a transportation infrastructure focused on road-building and private automobile use and has shaped land-use patterns throughout the country. The emphasis on motorized transportation has been associated with more driving, less physical activity, higher rates of obesity, higher rates of air pollution, and transportation injuries and fatalities. A *partial* accounting of the costs of health outcomes wholly or partly associated with transportation indicates that the costs could be as great as $400 billion annually. No one can know how much the costs could have been reduced if health had been integrated into the decision-making. Without a systematic assessment, the health-related effects and their costs to individuals and society are hidden or invisible products of transportation-related decisions.

Several approaches, methods, or tools could be used to incorporate aspects of health into decision-making, but HIA holds particular promise because of its applicability to a broad array of policies, programs, plans, and projects; its consideration of adverse and beneficial health effects; its ability to consider and incorporate various types of evidence; and its engagement of communities and stakeholders in a deliberative process. The following sections define and describe the elements of HIA, the challenges to its practice, and the approaches to advancing it and integrating it into today's decision-making processes.

DEFINING HEALTH IMPACT ASSESSMENT AND ITS ELEMENTS

On the basis of its review of HIA definitions, practice, published guidance, and peer-reviewed literature, the committee recommends the following technical definition of HIA, which is adapted from the definition of the International Association for Impact Assessment:

> HIA is a systematic process that uses an array of data sources and analytic methods and considers input from stakeholders to determine the potential effects of a proposed policy, plan, program, or project on the health of a population and the distribution of those effects within the population. HIA provides recommendations on monitoring and managing those effects.

The committee emphasizes that HIA is conducted to inform a decision-making process and is intended to be concluded and communicated in advance of a decision so that the information that it yields can be used to shape a final proposal in such a way that adverse effects are minimized and beneficial ones are optimized. The committee acknowledges that other assessment methods may share some features with HIA, but they do not meet the definition and description of HIA that the committee provides in the present report.

The committee found remarkable consistency regarding the basic elements that are generally included in descriptions of HIA, although they may be organized differently in the stages or steps that are outlined. The committee recommends a six-step framework as the clearest way to organize and describe the elements of HIA. The steps and their outputs are illustrated in Figure S-1; the committee's conclusions regarding each step are provided below.

Screening establishes the need for and value of conducting an HIA and is essential for high-quality HIA practice. The committee concludes that the following factors are the most important to consider in determining whether to conduct an HIA: the potential for substantial adverse or beneficial health effects or irreversible or catastrophic effects, even if the effects have a low likelihood; the ability of information from the HIA to alter a decision or help a decision-maker to discriminate among options; the possibility that a disproportionate burden of the health effects is placed on vulnerable populations; the existence of public concern or controversy regarding health effects of a proposal; the opportunity to incorporate health information into a decision-making process that may not otherwise include such information; and the ability of the HIA team to complete the assessment within the time and with the resources available.

Scoping identifies the populations that might be affected, determines which health effects will be evaluated in the HIA, identifies research questions and develops plans to address them, identifies the data and methods to be used and alternatives to be assessed, and establishes the HIA team and a plan for stakeholder participation throughout the HIA process. The credibility of an HIA and its relevance to the decision-making process rest on a systematic evaluation of the full array of potential effects—risks, benefits, and tradeoffs—rather than on a narrow consideration of a subset of issues predetermined by a team's research interests or regulatory requirements. However, to ensure judicious use of resources, the HIA should ultimately focus on the health effects of greatest potential importance. The committee notes that it is appropriate to include issues that are the subject of community concern even if they appear unlikely to be substantiated by further analysis; such an analysis can provide reassurance to communities even if the eventual conclusions do not support their concerns.

Assessment is a process that involves describing the baseline health status of the affected populations and then characterizing the expected effects on health (and its determinants) of the proposal and each alternative under consideration relative to the baseline and each other. In light of the various policies, programs, plans, and projects that are the subject of HIAs, a broad array of data and analytic methods are used to evaluate the potential effects. Often, complete information is not available, and expert judgment plays an important role in the HIA. Whatever approach is taken, an explicit statement of data sources, methods, assumptions, and uncertainty is essential. The committee notes that uncertainty does not negate the value of information. Even when the evidence of an effect is uncertain, describing the potential causal pathways that are based on a reasonable interpretation of available data and expert judgment can help to establish a framework for monitoring and managing any impacts that might occur as the proposal is implemented.

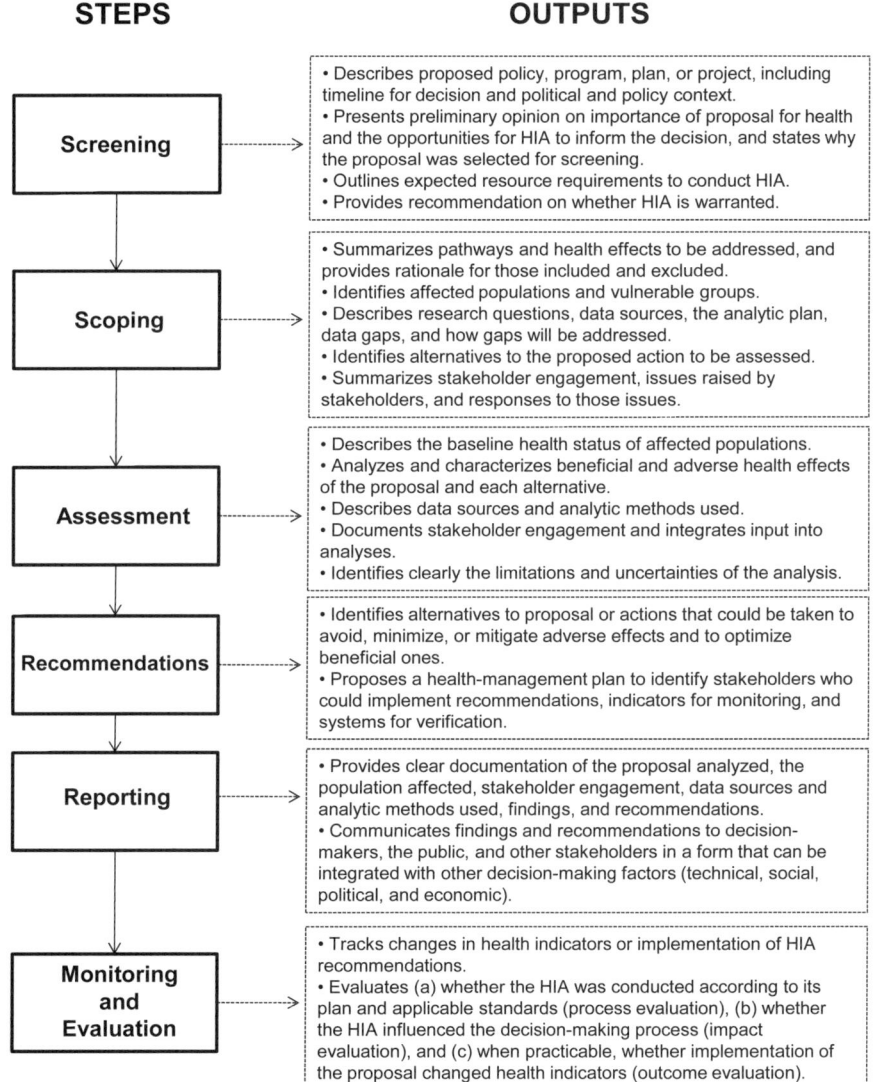

FIGURE S-1 Framework for HIA, illustrating steps and outputs.

Recommendations identify alternatives to the proposal or specific actions that could be taken to avoid, minimize, or mitigate adverse effects or to take advantage of opportunities for a proposal to improve health. Relatively little attention has been paid to the formulation of effective, actionable recommendations, and the committee offers three points for consideration. First, community input is essential for proposals that could have localized effects because it helps to ensure that specific aspects of living conditions and community design that may not be obvious to outside researchers are considered, and it maximizes the probability that the affected community will accept the conclusions and recommendations of the assessment. Second, recommendations are effective only if they are adopted by a decision-maker and implemented. The chances that the recommendations are adopted and implemented will increase if measures are drafted to address identified public-health risks; recognize feasibility issues, practical challenges, and other concerns possibly raised by the decision-maker during the HIA process; and fulfill the requirements of the legal and policy framework governing the decision. Third, recommendations should include the elements of a health-management plan that identifies appropriate indicators for monitoring, an entity with authority or ability to implement each measure, and a mechanism for verifying implementation and compliance. In practice, the HIA team will be asking a decision-maker to consider the findings and recommendations; ultimately, the decision-maker must balance health considerations with the many other technical, social, political, and economic concerns that bear on the proposal.

Reporting is the communication of findings and recommendations to decision-makers, the public, and other stakeholders. At present, there is little uniformity in the content of an HIA report. The committee recommends that, at a minimum, the written HIA report describe the proposed action or policy and alternatives that are the subject of the HIA, document the data sources and analytic methods used, identify the people consulted during the HIA process, and provide a clear, concise, and easily understood description of the process, findings, and recommendations. Furthermore, the reports should be made publicly available. A well-designed dissemination strategy is critical for the success of an HIA, and continuing efforts to inform decision-makers and stakeholders of the findings and recommendations are essential. However, efforts to support health-based recommendations must be carefully distinguished from biased efforts to promote a specific alternative on the basis of a skewed comparison of favorable and unfavorable aspects of a proposal or a predetermined political agenda. Undue bias in an HIA will likely compromise its credibility and efficacy.

Monitoring and evaluation can be characterized by several activities. Monitoring can consist of tracking the adoption and implementation of HIA recommendations or tracking changes in health indicators (health outcomes or health determinants) as a new policy, program, plan, or project is implemented. Evaluation can be process evaluation (evaluation of whether the HIA was conducted according to its plan of action and applicable standards), impact evaluation (evaluation of whether the HIA influenced the decision-making process), or outcome evaluation (evaluation of whether implementation of the proposal changes health outcomes or health determinants). Few HIA evaluation data have been published in the United States or elsewhere, and it is not reasonable to expect that decision-makers will adopt HIA widely in the absence of evidence of its effectiveness and value. Consequently, the committee concludes that the lack of attention to evaluation is a barrier that will need to be overcome if HIA is to be advanced in the United States and notes that unbiased evaluation of its effectiveness and value will require participation of evaluators independent of the HIA team, stakeholders, decision-makers, and fiscal sponsors.

The committee emphasizes that the definitions and criteria recommended here should not be considered rigid requirements but rather reflect an ideal of practice. Given the broad array of applications and the resources and time available for HIA, deviations are expected, but they should be justified by a clear and well-articulated rationale. The committee also notes that HIA should not be assumed to be the best approach to every health-policy question but should be seen as part of a spectrum of public-health and policy-oriented approaches; the most appropriate will depend on the situation and decision-making context.

CHALLENGES AHEAD FOR HEALTH IMPACT ASSESSMENT

The committee identified several challenges for the successful emergence, development, and practice of HIA. Many are related to various aspects of HIA practice and are noted below with the committee's suggestions for possible resolutions.

Defining health and the boundaries for HIA. As noted above, there is a growing consensus that individual health and public health are shaped by genetic, behavioral, social, economic, and environmental factors. Therefore, the committee concludes that HIA practice should not be restricted by a narrow definition of health or restricted to any particular policy sector (for example, education, urban planning, or finance), level of government (federal, state, tribal, or local), type of proposal (policy, program, project, or plan), or specific health outcome or issue (for example, asthma or obesity). There is no evidence to suggest that HIA is more important, appropriate, or effective in any particular decision context. On the contrary, HIA may be useful in a broad array of decision contexts, including many decision types to which it has not yet been applied. Accordingly, HIA should be focused on applications that present the greatest opportunity to protect or promote health and to raise awareness of the health consequences of decision-making. Because there are few legal mandates for HIA in the United States, it is most often conducted as a voluntary practice. As such, it will be difficult to ensure that decisions that could have the greatest impact on health are selected for evaluation. Thus, the current ad hoc approach to conducting HIA may result in less useful applications. The committee concludes that any future policies, standards, or regulations for HIA should include explicit criteria for identifying and screening candidate decisions and rules for providing oversight for the HIA process; such criteria and rules would promote the utility, validity, and sustainability of HIA practice.

Balancing the need to provide timely, valid information with the realities of varying data quality. HIA must provide evidence-based findings and recommendations within the practical realities and timelines of the decision-making process; however, HIA practitioners often face substantial challenges regarding data availability and quality.[1] The committee offers three strategies to maximize the validity of findings and recommendations in light of data constraints. First, one should consider diverse types of evidence and use expertise from multiple disciplines. Second, one should critically evaluate the data quality and select the evidence and analytic methods that are the strongest from among those available for a particular decision and context. There are no uniform standards for evaluating all potential evidence used in HIA, given the diverse applications and heterogeneity of data; in the future, criteria for data quality could be developed to characterize the relative strength of evidence and the nature and magnitude of uncertainties. Third, a strategy for assessing, acknowledging, and managing uncertainties is essential for ensuring the credibility of HIA findings and recommendations.

Producing quantitative estimates of health effects. Many expect HIA to produce quantitative estimates of health effects. Quantitative estimates of health effects have a number of desirable properties: they provide an indication of the magnitude of health effects, they can be easily compared with existing numerical criteria or thresholds that define the significance of particular effects, they allow one to make more direct comparisons among alternatives, and they provide inputs for economic valuation. They can be produced when there has been sufficient empirical research on relationships between particular determinants and health outcomes. Relying exclusively on quantitative estimation in HIA, however, presents some drawbacks. First, quantification has high information requirements. Given the breadth of health effects potentially considered in HIA, the sparse data available to support quantitative approaches, and the variability in practitioner capacity, it would be challenging or impossible for all HIAs to predict all potentially important health effects quantitatively. Second, because quantification can be resource-intensive, it may require more time than is practical, given the timeline for decision-making. Third, quantitative estimates may create an unwarranted impression of objectivity, precision, and importance and lead a reader to give credence to quantified results even if assumptions used in the analysis were based on subjective choices. Overall, however, quantitative estimates of health effects have value and should be provided when the data and resources allow and when they are responsive to decision-makers' and stakeholders' information needs.

Synthesizing conclusions on dissimilar health effects. Given that HIA analyzes multiple health effects, a practical challenge is synthesizing and presenting results on dissimilar health effects in a manner that is intelligible and useful to

---

[1] In this report, the term *HIA practitioners* refers to the people conducting the HIA.

decision-makers and stakeholders. Although summary measures have not been commonly used in HIA practice, they can be used to translate estimated effects on disparate health outcomes into a single comparable unit, such as quality-adjusted life years, disability-adjusted life years, and healthy-years equivalent. Calculating summary measures, however, requires assumptions and weighting schemes that need to be recognized and explained, and summary measures may not allow the integration of all health effects. Therefore, if summary measures are used, the committee recommends that effects—including those excluded from the summary measure—be described and characterized separately with regard to magnitude and significance in a way that allows users to judge their cumulative nature. The relative value of dissimilar health effects can then be considered explicitly or implicitly in the decision-making process.

Engaging stakeholders. Ensuring that stakeholders are able to participate effectively in the HIA process is widely described as an essential element of practice, although stakeholders often are not engaged or are only minimally engaged in the process. That discrepancy can be attributed to several factors, including the time and resources available; the methods, guidance, and standards used to conduct HIA; the importance that the practitioner or sponsor places on stakeholder participation; and a view that stakeholder participation may interfere with or impede progress. However, stakeholder participation is critical for the quality and effectiveness of the HIA. It helps to identify important issues; focus the HIA scope; highlight local conditions, health issues, and potential effects that may not be obvious to practitioners from outside the community; and ensure that recommendations are realistic and practical. Thus, whenever possible, strategies for stakeholder participation should extend beyond some minimal effort and address barriers and challenges to participation.

Ensuring the quality and credibility of HIA. Although HIA is different from primary scientific research, the committee concludes that several aspects of the HIA process could benefit from peer review. Peer review could highlight overlooked issues, identify opportunities to improve data or methods, and increase the legitimacy of conclusions and their acceptance and utility in the decision-making process. A formal peer-review process would need to overcome several obstacles, such as the possible difficulties in assembling the multidisciplinary team that would be needed to perform the review, the substantial delays that could occur in the process, and the current lack of agreed-on evaluation criteria. However, HIA is often conducted on proposals that are contested among polarized and disparate interests and stakeholders, and accusations of bias can arise. Independent peer review could help to ensure that the process by which HIA is conducted and the conclusions and recommendations produced are as impartial, credible, and scientifically valid as possible. The committee notes, however, that some flexibility in the peer-review process would be necessary particularly for cases in which an HIA must be completed rapidly to be relevant to the decision that it is intended to inform.

Managing expectations. HIA clearly is intended to inform decisions and ultimately to shape policy, programs, plans, and projects so that adverse health effects are minimized and potential health benefits are optimized. The hope is that identifying valid information on a decision's harms or benefits to health will motivate decision-makers to take protective actions. However, health typically is only one factor in the decision-making process; practical factors—such as cost, feasibility, and regulatory authority—also play a prominent role. And improved knowledge alone cannot necessarily change the ideology, interests, and attitudes of decision-makers. Thus, it is not reasonable to consider HIA successful only if it changes decisions. Furthermore, looking at HIA only as a mechanism for advocacy will compromise the support for and legitimacy of the practice.

Integrating HIA into environmental impact assessment (EIA). The U.S. National Environmental Policy Act (NEPA) and some related state laws explicitly require the identification and analysis of health effects when EIA is conducted. EIA, however, has traditionally included at most only a cursory analysis of health effects. Some argue that health analysis should be integrated into EIA because NEPA and related state laws provide a mechanism for achieving the same substantive goals as HIA. Others contend that EIA has become too rigid to accommodate a comprehensive health analysis and that attention should be focused on the independent practice of HIA. The committee emphasizes that the appropriate assessment of direct, indirect, and cumulative health effects in EIA under NEPA is a matter of law and not discretion, and recent efforts have successfully integrated the HIA framework into EIA. Thus, where legal standards under NEPA or applicable state EIA laws require an integrated analysis of health effects, one should be conducted with the same procedures that would be used to assess any other required factor. Because the steps and approaches of HIA and EIA are compatible, HIA offers an appropriate way to meet the requirement for health analysis under NEPA and related state laws. Although there are some substantive challenges to overcome, the committee concludes that improving the integration of health into EIA practice under NEPA and related state laws is needed and would advance the goal of improving public health.

ADVANCING HEALTH IMPACT ASSESSMENT

Substantial improvements in public health will require a focused effort to recognize and address the health consequences of decisions made at all levels and in all sectors of government. As noted, HIA is a particularly promising approach for integrating health implications into decision-making. International experience and the limited (but growing) experience in the United States provide important clues as to what is needed most to advance HIA.

Societal awareness of and education in HIA. First, the common belief that our health depends only on genetic predisposition, health care, and personal choice is impeding the improvement of public health. Policy-makers and the public need to be educated in the many factors that can affect health, the importance of considering them in all decision-making, and the role that HIA can play in the decision-making process. An education campaign will be necessary to secure the resources that will be needed for the development of HIA practice. Second, few U.S. academic institutions offer formal education in HIA. Consequently, there are few professionally trained HIA practitioners in the country, and there is little agreement among them as to what constitutes good practice. High-quality education and training will be vital for the advancement of HIA in the United States. Third, continuing education of HIA professionals, policymakers, and the public will be important for improving the quality of HIA practice in this country. The committee notes that a professional association or society could facilitate continuing education and develop, monitor, and facilitate standards of professional education and practice in HIA.

Structures and policies to support HIA. First, substantial interagency collaboration at the local, state, and federal levels is necessary to conduct HIA of policies, programs, plans, and projects, especially those emanating from nonhealth sectors, such as transportation, finance, urban planning, education, and agriculture. Such collaboration is essential, given the resource-constrained environments in which makers of public policy and other officials often work. The committee offers several suggestions for promoting interagency collaboration in the present report. Second, systematic use of HIA ultimately will depend on the adoption of policies and legal mandates to integrate health considerations into decision-making. As noted above, NEPA requires the analysis of health effects when EIA is conducted, but the spirit of the requirement needs to be reinvigorated and strengthened. Explicit guidance demonstrating how health considerations could be incorporated into NEPA would be beneficial. The committee emphasizes that policies and legislation outside the context of NEPA will most likely be needed to facilitate the use of HIA.

Research on and scholarship in HIA. First, few evaluations of HIA effectiveness have been conducted in the United States, especially because it has emerged so recently. Because conducting HIA will probably require the investment of substantial public and private resources, research is needed to document HIA practices and their effectiveness in influencing decision-making processes and promoting public health. Second, the quality of HIA could be substantially improved if there were better evidence on the relationship of "distal" factors to health outcomes. For example, research on how health is affected by federal, state, and local policies and actions traditionally considered to be unrelated to health—such as transportation, agriculture, education, housing, financial, and immigration policies—would be extremely beneficial.

The recognition that health is affected by much more than medical care, personal choice and behavior, and genetic predisposition is fundamental for the development and implementation of strategies to improve public health. However, the mere promulgation of a legal requirement to consider health would most likely not result in the health improvements that the United States needs. A tool, method, or approach is needed to facilitate the integration of health into decision-making. HIA is particularly promising in light of its broad applicability, its focus on adverse and beneficial health effects, its ability to incorporate various types of evidence, and its emphasis on stakeholder participation.

1

Introduction

There is growing evidence that our social, economic, and physical environments affect public health. Thus, our health is affected by how buildings and communities are designed, where roadways are located, and what economic, agricultural, and educational policies and programs are implemented. Health can no longer be seen solely as the result of personal choice and behavior. The task of integrating health considerations into such a breadth of activities is potentially daunting. However, a new field—health impact assessment (HIA)—can assist decision-makers in examining the potential health effects of proposed projects, programs, plans, and policies. It has gained momentum internationally, although it is not yet widely used in the United States. Some attribute the difference to the absence of a uniform framework and guidance for conducting such assessments. Given the potential of HIA to improve public health, the Robert Wood Johnson Foundation (RWJF), the National Institute of Environmental Health Sciences (NIEHS), the California Endowment, and the Centers for Disease Control and Prevention (CDC) asked the National Research Council (NRC) to develop a framework, terminology, and guidance for conducting HIA. As a result of that request, NRC convened the Committee on Health Impact Assessment, which prepared this report.

HEALTH IMPACT ASSESSMENT

The idea that many factors outside the traditional health field affect public health is not new. In fact, the decrease in mortality from infectious disease in the 19th and 20th centuries and the increase in life expectancy are attributed more to such factors as better nutrition, housing, and sanitation than to advances in medicine (McKeown 1979). Studies have demonstrated the relatively small influence of the medical practice on public health as opposed to the substantial effect of living conditions (Kemm and Parry 2004). Accordingly, many have recognized that improvements in public health will occur only if health considerations are factored into projects, programs, plans, and policies in non-health-related sectors, such as transportation, housing, agriculture, and education (Kemm and Parry 2004; Cole and Fielding 2007).

Given the studies of the determinants of public health, a new field, HIA, arose in the 1980s and 1990s. The most commonly cited definition of HIA was provided in what is known as the Gothenburg consensus paper:

> A combination of procedures, methods and tools by which a policy, programme or project may be judged as to its potential effects on the health of a population, and the distribution of those effects within the population (WHO 1999, p. 4).

Other definitions have arisen over the decades, and several examples are provided in Table 1-1. As shown, HIA has been defined in various ways and described by such terms as *method*, *process*, *approach*, *tool*, and *framework*. Diverse practices have been associated with HIA, and that diversity has been attributed somewhat to how health has been defined (or not defined) by the various governments and organizations that use HIA. Parry and Kemm (2004), however, asserted that the essential features of HIA are predicting the consequences of various options and educating and assisting decision-makers.

The International Experience

HIA has been used throughout the world to evaluate the potential health consequences of various projects, programs, plans, and policies (see Appendix A for discussion of the international experience in implementing HIA). Europe and such countries as Canada, Australia, and Thailand—and states, provinces, and territories in these countries—have used various approaches to introducing and promoting HIA. Some have integrated it into existing environmental-assessment frameworks or practices, and others have established it as a stand-alone or distinct process. Some have tried to legislate its use, and others have relied on voluntary processes in which various degrees of government support and resources are provided. Each country's experience offers different perspectives and lessons to be learned. For example, although the experience in a few countries has suggested that legislation is needed to provide an impetus for conducting HIA, the experience in many other countries has emphasized that legislative requirements alone are not sufficient to ensure its consistent implementation. Education, training, and resources appear to be critical to the success of its use, and engaging traditionally non-health-related sectors and agencies and heightening awareness of HIA also appear to be key.

International organizations have contributed to the development and evolution of HIA. Over the last few decades, the World Health Organization has supported the development and use of HIA through declarations, initiatives,

TABLE 1-1 Selected Definitions of Health Impact Assessment[a]

Definition	Reference
"Any *combination of procedures or methods* by which a proposed policy or program may be judged as to the effects it may have on the health of a population."	Frankish et al. 1996
"A *methodology* which enables the identification, prediction and evaluation of the likely changes in health risk, both positive and negative, (single or collective), of a policy, programme, plan or development action on a defined population. These changes may be direct and immediate, or indirect and delayed."	British Medical Association 1998, p. 39
"The *estimation* of the effects of a specified action on the health of a defined population."	Scott-Samuel 1998, p. 704
"A *method* of evaluating the likely effects of policies, initiatives and activities on health at a population level and helping to develop recommendations to maximise health gain and minimise health risks. It offers a framework within which to consider, and influence, the broad determinants of health."	Scottish Office Department of Health 1999, Section 98
"A *means* of evidence based policy making for improvement in health. It is a combination of methods whose aim is to assess the health consequences to a population of a policy, project, or programme that does not necessarily have health as its primary objective."	Scott Samuel 1997 in Lock 2000, p. 1395
"A *multidisciplinary process* within which a range of evidence about the health effects of a proposal is considered in a structured framework…based on a broad model of health, which proposes that economic, political, social, psychological, and environmental factors determine population health."	Grant et al. 2001, p. 1
"A *developing approach* that can help to identify and consider the potential—or actual—health impacts of a proposal on a population. Its primary output is a set of evidence-based recommendations geared to informing the decision making process."	Taylor and Quigley 2002, p. 2-3
"A *structured framework* to map the full range of health consequences of any proposal, whether these are negative or positive. It helps clarify the expected health implications of a given action, and of any alternatives being considered, for the population groups affected by the proposals. It allows health to be considered early in the process of policy development and so helps ensure that health impacts are not overlooked."	WHO 2002, p. 2

(Continued)

TABLE 1-1 Continued

Definition	Reference
"A *combination of procedures, methods and tools* that systematically judges the potential, and sometimes unintended, effects of a policy, plan, programme or project on the health of a population and the distribution of those effects within the population. HIA identifies appropriate actions to manage those effects."	Quigley et al. 2006, p. 1
"A *combination of procedures, methods, and tools* to assess the potential health impacts of a project on nearby populations, and to recommend mitigation measures. HIA addresses both negative and positive aspects of health. HIA will also try to identify benefits to health that may be enhanced."	IFC 2009, p. 4

[a]Key phrases have been highlighted in the definitions to indicate the various ways that HIA has been defined.
Sources: Krieger et al. 2003; Kemm and Parry 2004.

conferences, workshops, and networks (Cole and Fielding 2007; Forsyth et al. 2010). Its work was driven initially by the need to incorporate HIA into environmental assessments of water-management projects but soon broadened to encourage the use of HIA to define healthy public policies. Multilateral development banks and the International Finance Corporation have also contributed to the development of HIA; many have now adopted standards that include requirements to conduct HIA for projects submitted for funding (IFC 2009; Krieger et al. 2010; Harris-Roxas and Harris 2011).

Many countries and organizations have developed their own guidance on conducting HIA (for example, B.C. Ministry of Health 1994; Fehr 1999; NHS 2000; enHealth 2001; Abrahams et al. 2004; PHAC 2005; Quigley et al. 2006; Harris et al. 2007; IFC 2009; Metcalfe et al. 2009). Regardless of the similarity of the guidance, some have observed that no consistent approach or methods have been used (Kemm 2007; Bhatia 2010). Others have concluded that the criteria for initiating, conducting, and completing HIA need to be clarified (Krieger et al. 2003) and that terminology needs to be standardized (Kemm and Parry 2004). After reviewing numerous examples of HIA, Parry and Kemm (2004, p. 417) concluded that improvements are needed "in terms of methodological techniques and practical application if [HIA] is to truly fulfill its promise and become a useful adjunct to decision making."

Health Impact Assessment in the United States

In the United States, HIA as a practice independent of environmental or other regulatory impact assessment was first used in San Francisco in 1999 to evaluate a policy to increase the minimum wage (Bhatia and Katz 2001). Although not widely or commonly practiced, HIA has been used in all levels of government and across the country to evaluate health impacts of proposed projects, policies, plans, and programs. Much of the activity, however, has been centered on local communities, has focused on policies and programs associated with land-use, housing, and transportation planning, and has been sponsored by local public-health and planning agencies, nonprofit organizations, and academic institutions. Several academic institutions—notably the University of California, Berkeley and the University of California, Los Angeles—have helped to advance HIA at the local level by providing training and technical assistance and by developing methods and approaches for conducting HIA.

At the state level, Washington and Massachusetts have passed legislation to support HIA, and several other states—including California, Maryland, Minnesota, and West Virginia—have proposed legislation. Even without legislation, several states—such as Hawaii, Alaska, California, Wisconsin, and Oregon—have been conducting and using HIA to evaluate proposed projects, programs, plans, and policies.

At the federal level, the use of HIA has been largely in the context of implementing the National Environmental Policy Act (NEPA), which requires federal agencies to evaluate the health effects of proposed federal actions [42 U.S.C §§ 4321-4347]. However, the analysis of human health effects has historically been minimized in assessments conducted under NEPA. Several factors—including the lack of focus of early legal claims on human health, misinterpretation of case law, and the lack of involvement of traditionally health-related municipal, state, tribal, or federal agencies in the NEPA process—contributed to the de-emphasis of human health effects. That situation has changed recently with work conducted by native Alaskans to incorporate health, social, and cultural effects into NEPA documents for oil- and gas-leasing programs and leasing sales (BLM 2007; MMS 2007a,b; EPA 2009). That activity has focused attention on and promoted interest in HIA in various federal agencies (see Appendix A for further details on the HIA experience in the United States).

THE COMMITTEE'S TASK AND APPROACH

The committee that was convened in response to the request by RWJF, NIEHS, the California Endowment, and CDC includes experts in HIA, environmental impact assessment, public health, epidemiology, urban planning, social sciences, economics, and decision and risk analysis (see Appendix B for biographies of the committee members). The committee was asked specifically to develop a framework, terminology, and guidance for conducting HIA of proposed policies, programs, and projects at federal, state, tribal, and local levels, including the private sector. The committee was to assess the value and potential value of such assessments; the impediments and countervailing factors that have limited the practice of HIA to date; the circumstances and criteria for conducting HIA; the concepts, tools, and information required; and the types, structure, and content of HIA. On the basis of those considerations, the committee was to develop a systematic conceptual framework and approach for improving the assessment of health impacts in the United States (see Appendix C for the committee's statement of task).

To accomplish its task, the committee held five meetings. During the first three, public sessions were held in which the committee heard presentations by the sponsors and invited speakers in federal, state, and tribal government; academe; professional associations; nonprofit organizations; and consulting firms. The committee reviewed numerous publications on HIA and considered the experience of various countries and organizations in implementing HIA. A summary of the committee's review of HIA experience is provided in Appendix A. The committee's consideration of the literature and the HIA experience shaped its conclusions and recommendations for the framework and guidance that it offers here.

The committee notes that it was given a broad task, that is, to develop a framework and guidance for HIA applicable in all contexts. Therefore, the committee had to develop a flexible framework that is amenable to all types of HIA and could not simply provide a cookbook or technical manual on HIA. The committee, however, has provided extensive reference lists that should help to guide the reader with regard to specific assessments. Furthermore, the committee recognizes that HIA exists on a spectrum of impact assessment and planning tools that have been used for decades. However, the committee's focus was on developing a framework and guidance for HIA, not on comparing and contrasting all possible approaches and tools that are available. Similarly, although the committee reviewed the international and U.S. experience with HIA, it did not thoroughly examine and compare all types of HIAs that have been conducted or determine their impact and how the information has been used on release of the HIA. Finally, the committee uses various terms throughout the report, many of which are defined in the glossary (see Appendix D). The committee notes that it uses the term *public health* in this report in the broadest sense possible, that is, generally the health of the public. Implicit in the concept of public health used by the committee is the idea that health is affected by a wide array of factors that range from the societal to the biologic.

ORGANIZATION OF REPORT

The committee's report is organized into five chapters and six appendixes. Chapter 2 discusses the rationale for conducting HIA and the key role that it can play in improving public health and reducing health disparities. Chapter 3 outlines the elements of the HIA process (that is, the framework), describes the current variability, and highlights features that the committee finds are critical for any HIA. Chapter 4 provides the committee's suggestions for best practices for conducting HIA, and Chapter 5 discusses what is needed for advancing HIA. The review of HIA experience, the committee biographies, the statement of task, a glossary of commonly used terms, and a discussion of the analysis of health effects under NEPA are provided in appendixes.

REFERENCES

Abrahams, D., A. Pennington, A. Scott-Samuel, C. Doyle, O. Metcalfe, L. den Broeder, F. Haigh, O. Mekel, and R. Fehr. 2004. European Policy Health Impact Assessment: A Guide, University of Liverpool, England; RIVM, Netherlands; Institute of Public Health, Ireland; loegd, Institute of Public Health, NRW Bielefeld, Germany. Prepared for the Health and Consumer Protection Directorate General, European Commission. May 2004 [online]. Available: http://ec.europa.eu/health/ph_projects/2001/monitoring/fp_monitoring_2001_a6_frep_11_en.pdf [accessed May 16, 2011].

B.C. Ministry of Health. 1994. Health Impact Assessment Toolkit: A Resource for Government Analysis. Population Health Resource Branch, Ministry of Health, Vancouver, British Columbia.

Bhatia, R. 2010. A Guide for Health Impact Assessment. California Department of Public Health. October 2010 [online]. Available: http://www.cdph.ca.gov/pubsforms/Guidelines/Documents/HIA%20Guide%20FINAL%202010-19-10.pdf [accessed Apr. 22, 2011].

Bhatia, R., and M. Katz. 2001. Estimation of health benefits accruing from a living wage ordinance. Am. J. Public Health 91(9):1398-1402.

BLM (Bureau of Land Management). 2007. Northeast National Petroleum Reserve-Alaska (NPR-A) Draft Supplemental Integrated Activity Plan/Environmental Impact Statement (IAP/EIS). U.S. Department of the Interior, the Bureau of Land Management [online]. Available: http://www.blm.gov/ak/st/en/prog/planning/npra_general/ne_npra/northeast_npr-a_draft.html [accessed Nov. 30, 2010].

British Medical Association. 1998. Health and Environmental Impact Assessment. London: Earthscan.

Cole, B.L., and J.E. Fielding. 2007. Health impact assessment: A tool to help policy makers understand health beyond health care. Annu. Rev. Public Health 28:393-412.

enHealth (enHealth Council). 2001. Health Impact Assessment Guidelines. Canberra: Commonwealth of Australia. September 2001 [online]. Available: http://www.health.gov.au/internet/main/publishing.nsf/content/35F0DC2C1791C3A2CA256F1900042D1F/$File/env_impact.pdf [accessed May 5, 2011].

EPA (U.S. Environmental Protection Agency). 2009. Red Dog Mine Extension Aqqaluk Project. Final Supplemental Environmental Impact Statement. Prepared for U.S. Environmental Protection Agency, Seattle, WA, by Tetra Tech, Inc., Anchorage, AK. October 2009 [online]. Available: http://www.reddogseis.com/Docs/Final/Front_Matter.pdf [accessed Nov. 30, 2010].

Fehr, R. 1999. Environmental health impact assessment: Evaluation of a 10 step model. Epidemiology 10(5):618-625.

Frankish, C.J., L.W. Green, P.A. Ratner, T. Chomik, and C. Larsen. 1996. Health Impact Assessment as a Tool for Population Health Promotion and Public Policy. Prepared for Public Health Agency of Canada, by Institute of Health Promotion Research, University of British Columbia [online]. Available: http://www.phac-aspc.gc.ca/ph-sp/impact-repercussions/index-eng.php [accessed Apr. 22, 2011].

Forsyth, A., C.S. Slotterback, and K. Krizek. 2010. Health impact assessment (HIA) for planners: What tools are useful? J. Plan. Lit. 24(3):231-245.

Grant, S., J.R. Wilkinson, and A. Learmonth. 2001. An Overview of Health Impact Assessment. Occasional Paper No. 1. Technical report. Northern and Yorkshire Public Health Observatory, Stockton on Tees, UK. May 2001 [online]. Available: http://dro.dur.ac.uk/5613/ [accessed May 9, 2011].

Harris, P., B. Harris-Roxas, E. Harris, and L. Kemp. 2007. Health Impact Assessment: A Practical Guide. Sidney, Australia: Centre for Health Equity Training, Research and Evaluation, the University of New South Wales. August 2007 [online]. Available: http://www.hiaconnect.edu.au/files/Health_Impact_Assessment_A_Practical_Guide.pdf [accessed May 9, 2011].

Harris-Roxas, B., and E. Harris. 2011. Differing forms, differing purposes: A typology of health impact assessment. Environ. Impact Assess. Rev. 31(4):396-403.

IFC (International Finance Corporation). 2009. Introduction to Health Impact Assessment. Washington, DC: World Bank [online]. Available: http://www.ifc.org/ifcext/sustainability.nsf/AttachmentsByTitle/p_HealtheImpactAssessment/$FILE/HealthImpact.pdf [accessed May 5, 2011].

Kemm, J. 2007. What is HIA and why might it be useful? Pp. 3-13 in The Effectiveness of Health Impact Assessment: Scope and Limitations of Supporting Decision-Making in Europe, M. Wismar, J. Blau, K. Ernst, and J. Figueras, eds. Trowbridge, Wilts, UK: The Cromwell Press.

Kemm, J., and J. Parry. 2004. What is HIA? Introduction and overview. Pp. 1-13 in Health Impact Assessment: Concepts, Theory, Techniques, and Applications, J. Kemm, J. Parry, and S. Palmer, eds. Oxford: Oxford University Press.

Krieger, G.R., J. Utzinger, M.S. Winkler, M.J. Divall, S.D. Phillips, M.Z. Balge, and B.H. Singer. 2010. Barbarians at the gate: Storming the Gothenburg consensus. Lancet 375(9732):2129-2131.

Krieger, N., M. Northridge, S. Gruskin, M. Quinn, D. Kriebel, G. Davey Smith, M. Bassett, D.H. Rehkopf, and C. Miller. 2003. Assessing health impact assessment: Multidisciplinary and international perspectives. J. Epidemiol. Community Health 57(9): 659-662.

Lock, K. 2000. Health impact assessment. BMJ 320(7246):1395-1398.

McKeown, T. 1979. The Role of Medicine: Dream, Mirage, or Nemesis. Oxford, UK: Blackwell.

Metcalfe, O., C. Higgins, and T. Lavin. 2009. Health Impact Assessment Guidance. Institute of Public Health in Ireland [online]. Available: http://www.publichealth.ie/files/file/IPH%20HIA_0.pdf [accessed May 9, 2011].

MMS (Minerals Management Service). 2007a. Outer Continental Shelf Oil and Gas Leasing Program: 2007-2010. Final Environmental Impact Statement, Vol. 1. OCS EIS/EA MMS2007-003. U.S. Department of the Interior, Minerals Management Service, Herndon, VA. April 2007 [online]. Available: http://www.boemre.gov/5-year/2007-2012FEIS/Intro.pdf [accessed Nov. 30, 2010].

MMS (Minerals Management Service). 2007b. Chukchi Sea Planning Area Oil and Gas Sale 193 and Seismic Surveying Activities in the Chukchi Sea. Final Environmental Impact Statement. OCS EIS/EA MMS2007-026. U.S. Department of the Interior, Minerals Management Service, Alaska OCS Region [online]. Available: http://alaska.boemre.gov/ref/EIS%20EA/Chukchi_FEIS_193/feis_193.htm [accessed Nov. 30, 2010].

NHS (National Health Service). 2000. A Short Guide to Health Impact Assessment: Informing Healthy Decisions. NHS Executive, London [online]. Available: http://www.who.int/hia/examples/en/HIA_londonHealth.pdf [accessed May 9, 2011].

Parry, J., and J. Kemm. 2004. Future directions in HIA. Pp. 411-417 in Health Impact Assessment: Concepts, Theory, Techniques, and Applications, J. Kemm, J. Parry, and S. Palmer, eds. Oxford: Oxford University Press.

PHAC (Public Health Advisory Committee). 2005. A Guide to Health Impact Assessment: A Policy Tool for New Zealand, 2nd Ed. Wellington, New Zealand: PHAC [online]. Available: http://www.phac.health.govt.nz/moh.nsf/pagescm/764/$File/guidetohia.pdf [accessed May 9, 2011].

Quigley, R., L. den Broeder, P. Furu, A. Bond, B. Cave, and R. Bos. 2006. Health Impact Assessment: International Best Practice Principles. Special Publication Series No. 5. Fargo: International Association for Impact Assessment. September 2006 [online]. Available: http://www.iaia.org/publicdocuments/special-publications/SP5.pdf [accessed May 6, 2011].

Scott-Samuel, A. 1997. Assessing how public policy impacts on health. Healthlines 47 (Nov.):15-17.

Scott-Samuel, A. 1998. Health impact assessment—theory into practice. J. Epidemiol. Community Health 52(11):704-705.

Scottish Office Department of Health. 1999. Towards a Healthier Scotland: A White Paper on Health. Edinburgh: The Stationery Office.

Taylor, L., and R. Quigley. 2002. Health Impact Assessment: A Review of Reviews. National Health Service, Health Development Agency, London [online]. Available: http://www.nice.org.uk/niceMedia/documents/hia_review.pdf [accessed May 10, 2011].

WHO (World Health Organization). 1999. Health Impact Assessment: Main Concepts and Suggested Approaches-the Gothenburg Consensus Paper. Brussels: European Centre for Health Policy, WHO Regional Office for Europe.

WHO (World Health Organization). 2002. Health Impact Assessment: A Tool to Include Health on the Agenda of Other Sectors: Current Experience and Emerging Issues in the European Region. Technical Briefing, Regional Committee for Europe, 52nd Session, September, 16-19, 2002, Copenhagen [online]. Available: http://www.euro.who.int/__data/assets/pdf_file/0004/117049/ebd3.pdf [accessed Apr. 22, 2011].

2

Why We Need Health-Informed Policies and Decision-Making

On the basis of the most recent data from the World Health Organization, the United States ranks 32nd in the world in life expectancy—behind such countries as Japan, Australia, Italy, Greece, Iceland, Malta, and Luxembourg—despite ranking third in total expenditures on health care as a percentage of gross domestic product (GDP) (WHO 2010). Clearly, the United States still faces important challenges to promoting health and enhancing quality of life. For example, chronic diseases, many of which are preventable, account for more than 50% of all deaths each year (King et al. 2008). Almost half of all adults have at least one chronic illness (Wu and Green 2000). Obesity, a major risk factor for numerous health conditions, has grown to epidemic proportions in the United States (Ogden et al. 2007, 2008): one-third of all adults and almost one-fifth of people 6-19 years old are obese. Improvement in health has been inconsistent, and major disparities in health associated with socioeconomic circumstances, race, and ethnicity persist (Williams et al. 2010).

Despite major medical advances and large health expenditures, many Americans are unable to achieve their full health potential; this affects not only the quality and duration of their lives but their ability to be engaged and productive members of society. Poor health also has important economic implications—for lost productivity and for the costs of diagnosing and treating chronic conditions. Those costs affect individuals, communities, and society at large (WHO 2001; Hammitt 2007; Mackenbach et al. 2007). For example, costs for medical care have mushroomed both in amount and as a portion of the U.S. GDP because of the increases in medical care itself, the increases in use of the health-care system, the aging of the population, and the higher rates of chronic diseases. Health-care spending accounted for 7% of the U.S. GDP in 1970 and 16% of it in 2008 (CMS 2011); it is projected to be close to 20% by 2019 (CMS 2010), and this projection does not take into account the substantial increases in morbidity and mortality that will result from the obesity and diabetes epidemics.

Diabetes alone accounted for $174 billion in health-care costs in the United States in 2007; diabetes incidence is expected to increase from 7 per 1,000 to 15 per 1,000 by 2050 and diabetes prevalence from 14% to 21% by 2050 and in some scenarios up to 33% (Boyle et al 2010). Thus, the consequences of not preventing chronic health conditions are large, not only in years of healthy life lost but in monetary costs.

There is growing recognition among scientists, communities, and policymakers that health is affected by an array of factors that operate on multiple levels and throughout a person's lifetime (Adler and Stewart 2010). Although the importance of access to and quality of health care is well recognized, prevention is key. Disease prevention and health promotion require addressing a much broader set of factors and policies that shape health-related behaviors in addition to trying to modify biologic processes specifically related to diseases. Efforts to improve early detection and treatment of diseases through improved access to high-quality medical care must be complemented by approaches that address the underlying or root causes of disease. The underlying causes include the factors that shape the conditions in which people are born, grow, live, work, and age, and the policies that affect them. Those factors and their implications for health have been highlighted in a number of recent reports (see, for example, WHO 2002; CSDH 2008; RWJF 2009).

The root causes that have been identified indicate that many policies or programs thought to be unrelated to health may have important health consequences. Indeed, it has been argued that major health problems, such as the obesity epidemic and its associated health and monetary costs, are essentially unintended consequences of various social and policy factors related, for example, to the mass production and distribution of energy-dense foods (Ledikwe et al. 2006; Mendoza et al. 2007; Wang et al. 2008) and the engineering of physical activity out of daily life through changes in how transportation is organized and how neighborhoods are designed and built (Gordon-Larsen et al. 2005; Li et al. 2008; Frank and Kavage 2009; Fitzhugh et al. 2010). Such policy and planning decisions have powerful implications for individual behaviors and public health. The prevention of today's major health problems requires understanding and intervention to affect the root causes of ill health and the policies that shape and affect the root causes. To address them effectively, a better understanding of the possible health consequences of proposed policies and planning decisions as they are being developed is needed so that adverse health effects can be anticipated and minimized and health benefits maximized.

In summary, the health implications of decisions need to be considered explicitly not only to prevent harm but to promote health. Indeed, it can be argued that major improvements in the health of the U.S. public cannot be achieved without attention to the root causes of ill health and to the policies and programs that affect them. Furthermore, many root causes of ill health are common to the entire U.S. population, so interventions that address them can have broad-based impacts that benefit both high-risk groups and the general public.

KNOWLEDGE OF ROOT CAUSES OF HEALTH CONSEQUENCES

Research has identified measurable health consequences that have a wide variety of fundamental or root causes. The causes investigated have included broadly defined socioeconomic circumstances (Lynch et al. 1996; Marmot et al. 2001; Adler et al. 2007), education (Backlund et al. 1999; Din-Dzietham et al. 2000; Fleishman 2005; Lleras-Muney 2005; Kawachi et al. 2010), work and work environments (Marmot and Theorell 1988; Ferrie et al. 1998; Frank and Cullen 2006; Gillen et al. 2007; Cummings and Kreiss 2008; Ferrie et al. 2008; Clougherty et al. 2010), and physical and social features of communities or neighborhoods (Roberts 1997; Clougherty et al. 2007; Diez-Roux and Mair 2010). For example, a large literature has shown that economic resources are strongly associated with many health outcomes. The relationship between economic resources and health is not limited to those living in poverty; rather, there is abundant evidence of a graded inverse relationship between income and mortality or morbidity from chronic diseases that extends well above the poverty level (Adler and Stewart 2010).

Higher educational attainment is related to better health, possibly through the consequences of education for income, occupational achievement, residential location, and such other factors as self-efficacy and sense of control (Kawachi et al. 2010). For example, research shows that a 30-year-old white male high-school graduate can expect to live an average of 10 years longer than a 30-year-old white male who has less than 9 years of education. In black men, the education-based difference in life expectancy is greater than 16 years (Crimmins and Saito 2001).

Work environments are also important predictors of health. The adverse health consequences of physical and chemical exposures at work—such as exposure to toxicants, noise, and heat—are well established (Rosenstock et al. 2005). Recent work has shown that psychosocial features of the work environment, such as control of the work process, are important risk factors for chronic diseases (Siegrist 1996; Belkic et al. 2004; Ostry et al. 2006; Schulte et al. 2007; Clougherty et al. 2010; Krieger 2010; Meyer et al. 2010). It has also been suggested that trends in occupation-related physical activity may contribute to the obesity epidemic (Church et al. 2011).

There is abundant evidence of the impact of environmental factors, such as air pollution, on the causation and acceleration of respiratory and cardiovascular diseases (Brook et al. 2004; Dominici et al. 2006; Pope and Dockery 2006). In recent years, a broad and growing scientific literature has documented associations of various aspects of the physical and social environments of neighborhoods with health-related behaviors, such as diet and physical activity; these findings highlight important implications for the prevention of obesity, diabetes, and other chronic diseases (Brisbon et al. 2005; Hannon et al. 2006; Sturm 2008; Franzini et al. 2009; Larson et al. 2009; Chen and Florax 2010; Truong et al. 2010). Transportation systems and the location of industrial land uses are related to health; for example, childhood asthma (Gauderman et al. 2005; Jerrett et al. 2008; Mann et al. 2010; Mar et al. 2010), birth outcomes (Salam et al. 2005; Ritz et al. 2007; Slama et al. 2007; Woodruff et al. 2008), and cardiovascular risk (Brook et al. 2010; Park et al. 2010) have all been shown to be associated with transportation and planning decisions that shape exposure to air pollution, including airborne particulate matter and toxic gases generated by traffic and other sources. Health can be affected by planning decisions that result in urban sprawl (Pohanka and Fitzgerald 2004); for example, social isolation created by living in suburban areas may have health consequences (Pohanka and Fitzgerald 2004), and increased use of cars for commuting can result in increases in airborne particulate matter and in sedentary behavior associated with greater time spent in cars (Friedman et al. 2001).

A broad array of social and economic policies—although less frequently investigated in empirical studies—is likely to have measurable health impacts. For example, policies related to taxation, income supplementation, or access to education clearly determine a person's economic resources and educational attainment, which have been shown to affect health. Policies that affect job variety, quality, and environments will affect health, and policies that affect the physical and social environments of communities may also have important health consequences (Dow et al. 2010). Examples include housing policies that affect the quality and location of housing developments; transportation policies that affect the quality and availability of public transportation; urban-planning policies and decisions that affect land use and street connectivity or the creation of new housing developments; policies related to the location of food stores, farmers markets, and other food services; policies that promote safety and social interactions between neighbors, such as those related to community policing, lighting, organization, and design of attractive public spaces; and economic-development and zoning policies that affect the location of businesses and industries.

The factors that affect health are also root causes of health disparities associated with socioeconomic status, race, or ethnicity. Those health disparities are pronounced and persistent and do not appear to be declining despite medical advances. It is apparent that reducing the disparities will require addressing the more fundamental causes. Moreover, socioeconomically disadvantaged groups and racial or ethnic minorities are already at a health disadvantage and are the ones most likely to be affected by unintended adverse health consequences of policies or planning decisions because of where they live, their lack of resources to buffer or compensate adverse effects, and their lack of political power to advocate for their health. Indeed, even if a policy or decision improves public health overall, disparities in health related to socioeconomic position, race, or ethnicity may persist (Schulz and Northridge 2004; Frohlich and Potvin 2008).

WHY ASSESS THE HEALTH CONSEQUENCES OF POLICIES, PROGRAMS, PROJECTS, AND PLANNING DECISIONS?

Systematic assessment of the health consequences of policy, program, project, and planning decisions is of major importance for protecting and promoting public health because it allows the people who are involved in the decision-making process to consider the health impacts with other factors. Decisions can then be modified to minimize adverse health consequences or to maximize health benefits. Failure to consider health consequences can result in unintended harm or in lost opportunities for health improvement and disease prevention. Below are examples that illustrate the implications of failure to consider health consequences of policies, programs, projects, or plans.

U.S. agricultural-assistance programs provide subsidies for commodity crops—such as corn, soybeans, wheat, and rice—to help to ensure that U.S. families have an affordable source of food, that crop prices are stable, and that farmers continue to farm. Fruits, vegetables, and nonwheat grains are not subsidized, so farmers may be less likely to grow them. Although the assistance programs are considered successful, some researchers argue that an unintended consequence of the subsidies is their contribution to the current obesity epidemic and other nutrition problems (Fields 2004; Tillotson 2004; Hawkes 2007; Drewnowski 2010). For example, products made from the few subsidized crops—including high-fructose corn syrup sweeteners, hydrogenated fats made from soybeans, and feed for cattle and pigs—may saturate the market; this in turn may lower the prices of fattening, nutrient-poor, and energy-dense foods, such as prepackaged snacks, ready-to-eat meals, and fast food. The cheaper foods can easily compete with higher-priced healthier foods, such as fruits and vegetables, and this can affect calorie intake and other dietary factors that have implications for various chronic conditions, such as obesity, diabetes, and metabolic syndrome (Ledikwe et al. 2006; Mendoza et al. 2007; Wang et al. 2008). Lower-income groups may also be disproportionately affected by the less expensive, less nutritious foods because a larger portion of their diets may consist of these foods. The health consequences of policies promoting the production of inexpensive, calorie-dense foods could thus be far-ranging but remain unknown in the absence of a systematic assessment.

A second example of a failure to anticipate the health effects of policy and planning decisions is apparent in examining the health effects of transportation infrastructure. The Interstate Highway Act of 1956 introduced the development of a transportation infrastructure that has had multiple implications for health, both favorable and unfavorable. Over the last several decades, the transportation infrastructure has focused on road-building, private automobiles, and transportation of goods and has resulted in "an unprecedented level of individual mobility and facilitated economic growth" (APHA 2010, p. 2). It has shaped land-use patterns throughout the United States and has had implications for air quality, toxic exposures, noise, traffic collisions, pedestrian injuries, and neighborhood physical and social features potentially linked to health (Frank et al. 2006).

Transportation accounts for 30% of U.S. energy demand, and in 2008, tailpipe emissions from motor vehicles and impacts from fuel production contributed an estimated $56 billion in health and related damages (NRC 2010).[1] The costs partly reflect transportation-investment decisions that are focused on maximizing the safety and efficiency of automobile use and have resulted in important efficiencies in motor-vehicle transportation. The decisions have also led to transportation systems that discourage pedestrian and bicycle travel because of sheer distances between destinations, lack of adequate infrastructure for pedestrian travel, and increased hazards associated with pedestrian traffic—for example, unsafe pedestrian crossings and absence of pedestrian routes that are separate and safe from motor vehicles (APHA 2010). Personal and societal costs of the transportation decisions include nearly 34,000 deaths in 2009 due to motor-vehicle collisions; more than 12% of the deaths were of pedestrians (NHTSA 2010). The emphasis on motorized transport has been associated with more driving (Ewing and Cervero 2001; Frank et al. 2007), less physical activity (Saelens et al. 2003; Frank et al. 2005, 2006; TRB 2005), higher rates of obesity (Ewing et al. 2003; Frank et al. 2004; Lopez 2004), and higher rates of air pollution (Frank et al. 2000; Frank and Engelke 2005; Frank et al. 2006). A partial accounting of costs associated with the health effects, shown in Table 2-1, totals about $400 billion in 2008.

There is evidence that adverse health effects associated with transportation disproportionately affect members of racial and ethnic minorities and those in lower socioeconomic strata and thus contribute to persistent racial, ethnic, and socioeconomic disparities in health (Houston et al. 2004; Apelberg et al. 2005; Ponce et al. 2005; Wu and Batterman. 2006; Chakraborty and Zandbergen 2007). In the absence of systematic assessment of health effects and their associated costs, the implications of transportation decisions for health and health inequities cannot be factored into the process of making decisions about transportation infrastructure. As a result, the health-related effects and their costs to individuals and society are hidden or invisible products of transportation-related decisions.

Both adverse and beneficial health effects of specific decisions may sometimes be manifested rapidly. A study of the health consequences of changes in transit systems during the 1996 Olympic Games in Atlanta documented beneficial health effects of decisions made primarily to reduce downtown traffic congestion. Efforts to reduce congestion included daily 24-hour public transportation, the addition of 1,000 buses to support park-and-ride transit in the city, local

---

[1]The estimate excludes costs associated with climate change and non-fuel impacts, such as accidents and health effects resulting from reduced exercise.

TABLE 2-1 Costs of Transportation-Related Health Outcomes, 2008

Outcome	U.S. dollars, billions[a]	Factors Included in Estimate
Obesity[b]	$142	• Health-care costs • Lost wages due to illness and disability • Lost future earnings due to premature death
Air pollution from traffic	$50-80	• Health-care costs • Premature death
Traffic crashes	$180	• Health-care costs • Lost wages • Property damage • Travel delay • Legal and administrative costs • Pain and suffering • Lost quality of life

[a]All cost estimates are adjusted to 2008 U.S. dollars.
[b]"A portion of these costs are attributable to auto-oriented transportation and land use development that inadvertently limit opportunities for physical activity and access to healthy food" (APHA 2010, p. 2).
Source: Adapted from APHA 2010, page 4. Reprinted with permission; copyright 2010, American Public Health Association.

business use of alternative work hours and telecommuting, closure of the downtown sector to private automobile travel, alteration of downtown delivery schedules, and public announcements of potential traffic and air-quality problems. Those actions resulted in substantial decreases in acute childhood asthma events that were reversed after the end of the Olympic Games and the resumption of usual traffic patterns (Friedman et al. 2001).

Similarly, the introduction of electronic toll collection (E-ZPass), which reduced idling and queuing by allowing cars to move more quickly through toll booths, had important favorable effects on birth outcomes. Currie and Walker (2011) compared birth outcomes among women who lived near toll booths where E-ZPass was introduced with birth outcomes among women who lived near busy roadways that were not close to E-ZPass tollbooths. The introduction of E-ZPass greatly reduced traffic congestion and motor-vehicle emissions in the vicinity of highway toll plazas. The reductions in motor vehicle emissions were associated with a 10.8% reduction in prematurity and an 11.8% reduction in low birth weight of infants born to women living within 2 km of E-ZPass toll booths (Currie and Walker 2011). Moreover, there is substantial evidence that the probability of living near highways is unequally distributed by race, ethnicity, and socioeconomic status; this suggests that the changes may not only improve birth outcomes but reduce racial and socioeconomic disparities in those outcomes

(Gunier et al. 2003; Green et al. 2004; Houston et al. 2004; Jacobsen et al. 2004; Ponce et al. 2005).

In the examples above, health was not the primary force driving the decision to implement a policy or program, but important health consequences were observed. Moreover, the actions had consequences not only for public health generally but for disparities in health given that many of the conditions are more common among specific racial, ethnic, and socioeconomic groups. Integrating health considerations in a systematic way into the planning of programs, policies, and projects is key to preventing poor health and improving and protecting public health. The failure to consider consequences has led and will lead to many unanticipated adverse health consequences that have human and economic implications. The examples also demonstrate the potential of identifying unexpected health-enhancing policy and program interventions that can contribute substantially in addressing major health problems.

In summary, growing scientific evidence of the links between health and many economic, social, and planning factors makes it imperative to evaluate the health implications of policies, programs, projects, and plans that affect the root causes. Health-informed decision-making is sorely needed. The systematic assessment of the health consequences of policies and planning decisions is of special importance for protecting the health of vulnerable groups and those already at a health disadvantage because of adverse social or economic circumstances. In addition, it is fundamental to eliminating health disparities by race, ethnicity, and socioeconomic circumstances.

WHY ASSESSMENTS ARE NOT BEING CONDUCTED

Scientific information on the importance of root causes is abundant and growing, but it is not being fully used in a practical sense—that is, by applying it to the daily decisions made at the local, state, tribal, or federal level to enhance health and reduce health disparities. There are a number of reasons why health effects may not be systematically incorporated into decisions regarding policies, programs, projects, or plans, including the following:

- The absence of a mandate or funding to address root causes of ill health or health disparities or to assess the health impacts of planned policies and decisions.
- The presence of structural and administrative barriers to collaboration among public-health, planning, and environmental-health professionals (Epstein et al. 2006).
- The mismatch and lack of coherence among governance structures— for example, planning decisions about land use are made under the jurisdictions of local townships, and public-health decisions are made at the level of a city, county, or state.

- The perception that health and health disparities are attributable only to individual characteristics and choices (Link and Phelan 1995).
- The absence of inclusive and participatory mechanisms and processes for systematically integrating planning, public health, and environmental-health promotion in decision-making.
- The failure to enforce existing regulations to assess health implications of policies, programs, projects, and plans—for example, the failure to capture health impacts adequately in the context of environmental impact assessments.

Given the potential to reduce harm and enhance health, it is imperative to overcome the barriers that have prevented the consideration of health in decision-making. Factoring health and health-related costs into decision-making is essential in confronting the nation's pressing health problems and enhancing public health.

WHAT ARE THE OPTIONS FOR ASSESSMENT?

Assessing the health consequence of policies, programs, projects, and plans is a challenge that will require an interdisciplinary approach—involving such disciplines as health, social sciences, economics, and policy—and the collaboration of scientists, policy-makers, and communities. Systematic processes for rigorously assessing health consequences are needed. Although numerous analytic and deliberative tools are being used to incorporate aspects of health into decisions, none fully provides all the necessary attributes.

Human health risk assessment has been used for decades to incorporate understanding of the health implications of exposures (often environmental) into the regulatory decision-making process. However, risk assessment as conventionally practiced generally focuses on individual chemicals or limited multichemical exposure scenarios and does not capture the array of factors described earlier in this chapter. Although it could be argued that risk assessment can be applied in a manner that addresses all dimensions of policy influences on health and that the recent move toward cumulative risk assessment recognizes the need to consider a wide array of chemical and nonchemical exposures (NRC 2009), risk assessment without a substantial redefinition of the field is unlikely to be applicable to the great variety of policies, programs, projects, and plans that could have health implications.[2] Moreover, traditional risk assessment tends to focus on adverse health effects rather than on beneficial *and* adverse effects. It also emphasizes quantitative outputs as the primary end points in most appli-

---

[2]The committee notes that cumulative impact assessment as defined in NRC (2009) is somewhat broader than cumulative risk assessment in that it captures a wider array of end points and includes more qualitative components than cumulative risk assessment. However, it is generally oriented more toward characterizing impacts and less toward informing specific interventions or decisions.

cations. Although risk assessments include qualitative elements—such as hazard identification—and involve qualitative descriptions in risk characterization, they are generally secondary to the quantitative elements, and outcomes that cannot be quantified are rarely decision-relevant. Even in the context of cumulative risk assessment, NRC (2009) emphasized the importance of retaining the key attributes of quantitative risk assessment. Finally, it rarely engages stakeholders and communities in a deliberative process. Thus, in spite of the well-established regulatory mechanisms for health risk assessment and its potential to be modified in the long term, it is unlikely that all the health consequences of policy and planning decisions could be appropriately captured by conventional risk assessment (and in some situations, a narrow application of risk assessment could lead to policy and planning decisions that are injurious to health).

Other tools used to incorporate health into decision-making include cost-benefit or cost-effectiveness analysis, which often uses outputs from health risk assessment and the costs of implementing control strategies or other interventions. Those analytic tools commonly use a decision-theory framework in which various interventions are considered and an optimal choice is made on the basis of the outputs of the analysis. However, they have limitations similar to those surrounding traditional risk assessment, including a focus on more analytic than deliberative aspects of decision-making and a lack of an obvious mechanism to include qualitative information and participation of stakeholders.

The existing tool that may be most closely aligned with the consideration of multilevel and root causes is life-cycle assessment (LCA) (Curran 1996; EPA 2006). LCA examines a process or product and characterizes the full array of its upstream and downstream implications, including effects on human health, ecosystems, and other end points of interest to decision-makers. LCA typically relies on a combination of quantitative and qualitative evidence to compare various approaches to achieve a goal. However, LCA is generally more focused on such applications as manufacturing or fuel-cycle analysis and consists of more generic characterizations rather than site-specific characterizations. Thus, LCA attempts to characterize typical situations often from a national or global perspective, whereas the types of policies and planning decisions in which health dimensions need to be considered are often local and have unique site-specific attributes that should be considered.

Because of the limitations of existing tools in their ability to evaluate the health consequences of an array of policies, programs, projects, and plans systematically, health impact assessment (HIA) is a tool that holds promise for scientists, communities, and policy-makers. By its very nature, HIA lies at the intersection of science, policy, and stakeholder and community engagement. It includes attributes of health risk assessment, cost-benefit analysis, and LCA but differs from them in important ways, including its applicability to a variety of policies, projects, programs, and plans; its consideration of beneficial and adverse health consequences; its ability to consider and incorporate different types of evidence; and its engagement of communities and stakeholders in a deliberative process. HIA offers a way to engage agencies and individuals that do not normally work together, may not share a common expertise and knowledge, and often have differing priorities, authority, and objectives. It seeks to correct the fundamental problem of failing to consider health at all in decision-making. The committee concludes that HIA is valuable even with a lack of perfect forecasting data and tools because it is better to consider potential health risks and benefits than to ignore them routinely.

The committee acknowledges that other assessment approaches may share some features with HIA, but they do not meet the definition and description of HIA that the committee provides in the present report. Those defining features are discussed in detail in the chapters that follow.

OTHER BENEFITS OF SYSTEMATIC ASSESSMENT OF HEALTH IMPACTS

The committee concluded that HIA has at least three important benefits in addition to the obvious implications for improved policy-making and promotion and protection of health that would result from the systematic assessment of the health consequences of policies, programs, projects, and plans:

- *Improving the evidence.* The conduct of systematic assessments of health impacts will explicitly identify data gaps and evidence needed to improve future assessments. It will stimulate policy-relevant scientific research more directly, whether to develop new empirical studies or to improve systematic evaluation and synthesis of existing evidence. In addition, systematic monitoring of the health consequences of policies or actions after they are implemented should provide valuable new data directly relevant to answering policy-relevant causal questions that often cannot be addressed with observational studies or randomized trials. For example, in the Oak-to-Ninth Development Project HIA, the University of California, Berkeley, Health Impact Group conducted an analysis to estimate the effect of project-generated traffic on the frequency of pedestrian-automobile collisions in Chinatown in Oakland, California (UCBHIG 2007). Critiques and discussion of the results of the HIA led to the development and validation of a predictive model for pedestrian collisions (Wier et al. 2009) that was used in a later HIA (Bhatia and Wernham 2008). The process of systematic assessment, critique, and refinements in the development of scientific evidence to inform decision-making is critical for the development of health assessments that inform decision-making effectively.
- *Raising awareness among policy-makers and the public.* The systematic assessment of the health consequences of policies and planning decisions will raise awareness among policy-makers and the public at large about the wide variety of factors that affect health. It can contribute to a more comprehensive understanding of the causes of illness and of the role of policies, programs, projects, and plans in shaping health outcomes, including strategies that are likely to make the most difference in improving health and in reducing health disparities.

The recognition that health is affected by much more than lifestyle choices, genetic predispositions, and medical care is fundamental in the development and implementation of the types of strategies that are needed to improve public health. For example, the development of systematic evidence has resulted in a growing evidence base that links food policies and food access to the obesity epidemic and associated chronic diseases; the knowledge of these associations has in turn begun to generate attention and action among policy-makers (NAGC 2010).

- *A new paradigm for productive collaborations.* The assessment of the health consequences of policy and planning decisions will provide opportunities for a new paradigm for productive collaborations. For example, the collaborations offer opportunities (1) for scientists to be more directly involved in the application of the science that they conduct to improve public health and to be made more aware of the type of evidence needed for policy decisions, (2) for identification of new data sources and designs needed to answer important scientific and policy-relevant questions, (3) for improved ability of policy-makers to consider health implications in making decisions and improved understanding of the links between policies and health, (4) for active participation of community members in decision-making and increased access to information on health consequences available through the assessment process, which can enhance their ability to advocate for health, and (5) for improved insights into the potential pathways through which proposed decisions are likely to affect the health of residents (see, for example, Arquette et al. 2002; Corburn 2005).

The collaborations hold great potential for enhancing society's ability to prevent disease and promote public health. Furthermore, the active engagement of representatives of communities whose health stands to be affected by proposed policies, programs, projects, and plans is an essential component of democratic decision-making. Public engagement may also enhance understanding of the pathways through which policies, programs, projects, and plans may affect health and could promote actions that contribute to the reduction of health disparities. For example, the engagement of community members in HIA may lead to greater awareness of the impact of community resources on health and result in actions to improve community environments. Finally, systematic assessment of health consequences will give community groups a practical mechanism for increasing accountability of policy-makers and developers in the public and private sectors.

CONCLUSIONS

As a society, we routinely make decisions and implement a variety of policies, programs, and strategies without knowledge of their health implications. But those actions could substantially affect the health of the population and health disparities. The health consequences can have economic and social costs, which can have multiplying and cumulative effects. Identifying the potential effects in advance is fundamental for disease prevention and could have important consequences for trends in diseases and for social inequalities in a wide variety of health outcomes.

By tackling issues that other policy-analysis tools do not systematically incorporate or address, HIA has both a more expansive vision and a number of barriers to overcome to be accepted as a decision-making tool. Thus, it holds great potential but also presents a number of challenges. The following chapters discuss the key elements of HIA, review the status of HIA, and propose ways to improve the quality and utility of HIA in the future.

REFERENCES

Adler, N., and J. Stewart. 2010. Health disparities across the lifespan: Meaning, methods, and mechanisms. Ann. NY Acad. Sci. 1186:5-23.

Adler, N., J. Stewart, S. Cohen, M. Cullen, A.D. Roux, W. Dow, G. Evans, I. Kawachi, M. Marmot, K. Matthews, B. McEwen, J. Schwartz, T. Seeman, and D. Williams. 2007. Reaching for a Healthier Life: Facts on Socioeconomic Status and Health in the U.S. The John D. and Catherine T. MacArthur Foundation Research Network on Socioeconomic Status and Health [online]. Available: http://www.macses.ucsf.edu/downloads/Reaching_for_a_Healthier_Life.pdf [accessed Jan. 10, 2011].

Apelberg, B.J., T.J. Buckley, and R.H. White. 2005. Socioeconomic and racial disparities in cancer risk from air toxics in Maryland. Environ. Health Perspect. 113(6):693-699.

APHA (American Public Health Association). 2010. Backgrounder: The Hidden Health Costs of Transportation. Prepared by Urban Design 4 Health, Inc. and the American Public Health Association, Washington, DC. March 2010 [online]. Available: http://www.apha.org/NR/rdonlyres/8CB9D85D-3592-4C0B-8557-C22E925F75A7/0/FINALHiddenHealthCostsLongNewBackCover.pdf [accessed Jan. 10, 2011].

Arquette, M., M. Cole, K. Cook, B. LaFrance, M. Peters, J. Ransom, E. Sargent, V. Smoke, and A. Stairs. 2002. Holistic risk-based environmental decision-making: A native perspective. Environ. Health Perspect. 110(suppl. 2):259-264.

Backlund, E., P.D. Sorlie, and N.J. Johnson. 1999. A comparison of the relationships of education and income with mortality: The National Longitudinal Mortality Study. Soc. Sci. Med. 49(10):1373-1384.

Belkic, K.L., P.A. Landsbergis, P.L. Schnall, and D. Baker. 2004. Is job strain a major source of cardiovascular disease risk? Scand. J. Work Environ. Health 30(2):85-128.

Bhatia, R., and A. Wernham. 2008. Integrating human health into environmental impact assessment: An unrealized opportunity for environmental health and justice. Environ. Health Perspect. 116(8): 991-1000.

Boyle, J.P., T.J. Thompson, E.W. Gregg, L.E. Barker, and D.F. Williams. 2010. Projection of the year 2050 burden of diabetes in the U.S. adult population: Dynamic modeling of incidence, mortality, and prediabetes prevalence. Popul. Health Metr. 8:29.

Brisbon, N., J. Plumb, R. Brawer, and D. Paxman. 2005. The asthma and obesity epidemics: The role played by the built environment—a public health perspective. J. Allergy Clin. Immunol. 115(5):1024-1028.

Brook, R.D., B. Franklin, W. Cascio, Y. Hong, G. Howard, M. Lipsett, R. Luepker, M. Mittleman, J. Samet, S.C. Smith, Jr., and I. Tager. 2004. Air pollution and cardiovascular disease: A statement for healthcare professionals from the Expert Panel on Population and Prevention Science of the American Heart Association. Circulation 109(21):2655-2671.

Brook, R.D., S. Rajagopalan, C.A. Pope III, J.R. Brook, A. Bhatnagar, A.V. Diez-Roux, F. Holguin, Y. Hong, R.V. Luepker, M.A. Mittleman, A. Peters, D. Siscovick, S.C. Smith, Jr., L. Whitsel, and J.D. Kaufman. 2010. Particulate matter air pollution and cardiovascular disease: An update to the scientific statement from the American Heart Association. Circulation 121(21):2331-2378.

Chakraborty, J., and P.A. Zandbergen. 2007. Children at risk: Measuring racial/ethnic disparities in potential exposure to air pollution at school and home. J. Epidemiol. Community Health 61(12):1074-1079.

Chen, S.E., and R.J. Florax. 2010. Zoning for health: The obesity epidemic and opportunities for local policy intervention. J. Nutr. 140(6):1181-1184.

Church, T.S., D.M. Thomas, C. Tudor-Locke, P.T. Katzmarzyk, C.P. Earnest, R.Q. Rodarte, C.K. Martin, S.N. Blair, and C. Bouchard. 2011. Trends over 5 decades in U.S. occupation-related physical activity and their associations with obesity. PLoS One. 6(5): e19657.

Clougherty, J.E., J.I. Levy, L.D. Kubzansky, P.B. Ryan, S.F. Suglia, M.J. Canner, and R.J. Wright. 2007. Synergistic effects of traffic-related air pollution and exposure to violence on urban asthma etiology. Environ. Health Perspect. 115(8):1140-1146.

Clougherty, J.E., K. Souza, and M.R. Cullen. 2010. Work and its role in shaping the social gradient in health. Ann. NY Acad. Sci. 1186:102-124.

Corburn, J. 2005. Street Science: Community Knowledge and Environmental Health Practice. Cambridge, MA: MIT Press.

Crimmins, E.M., and Y. Saito. 2001. Trends in healthy life expectancy in the United States 1970-1990: Gender, racial, and educational differences. Soc. Sci Med. 52(11):1629-1641.

CSDH (Commission on Social Determinants of Health). 2008. Closing the Gap in a Generation: Health Equity through Action on the Social Determinants of Health. Geneva: World Health Organization [online]. Available: http://www.who.int/social_ determinants/thecommission/finalreport/en/index.html [accessed May 11, 2011].

CMS (Centers for Medicare & Medicaid Services). 2010. National Health Expenditure Projections 2009-2019. Centers for Medicare & Medicaid Services, Baltimore, MD [online]. Available: https://www.cms.gov/NationalHealthExpendData/down loads/proj2009.pdf [accessed July 11, 2011].

CMS (Centers for Medicare & Medicaid Services). 2011. National Health Expenditure Web Tables. Centers for Medicare & Medicaid Services, Baltimore, MD [online]. Available: http://www.cms.gov/NationalHealthExpendData/downloads/tables.pdf [accessed June 27, 2011].

Cummings, K.J., and K. Kreiss. 2008. Contingent workers and contingent health: Risks of a modern economy. JAMA 299(4):448-450.

Curran, M.A. 1996. Environmental Life Cycle Assessment. New York, NY: McGraw-Hill.

Currie, J. and R. Walker. 2011. Traffic congestion and infant health: evidence from E-ZPass. American Econ. J. Appl. Econ. 3(1):65-90.

Diez Roux, A.V., and C. Mair. 2010. Neighborhoods and health. Ann. NY Acad. Sci. 186:125-145.

Din-Dzietham, R., D. Liao, A. Diez-Roux, F.J. Nieto, C. Paton, G. Howard, A. Brown, M. Carnethon, and H.A. Tyroler. 2000. Association of educational achievement with pulsatile arterial diameter change of the common carotid artery: The Atherosclerosis Risk in Communities (ARIC) Study, 1987-1992. Am. J. Epidemiol. 152(7):617-627.

Dominici, F., R.D. Peng, M.L. Bell, L. Pham, A. McDermott, S.L. Zeger, and J.M. Samet. 2006. Fine particulate air pollution and hospital admission for cardiovascular and respiratory diseases. JAMA 295(10):1127-1134.

Dow, W.H., R.F. Schoeni, N.E. Adler, and J. Stewart. 2010. Evaluating the evidence base: Policies and interventions to address socioeconomic status gradients in health. Ann. NY Acad. Sci. 1186:240-251.

Drewnowski, A. 2010. The cost of U.S. foods as related to their nutritive value. Am. J. Clin. Nutr. 92(5):1181-1188.

EPA (U.S. Environmental Protection Agency). 2006. Life Cycle Assessment: Principles and Practice. U.S. Environmental Protection Agency [online]. Available: www.epa.gov/nrmrl/lcaccess/pdfs/600r06060.pdf [accessed June 23, 2011].

Epstein, L.H., S. Raja, S.S. Gold, R.A. Paluch, Y. Pak, and J.N. Roemmich. 2006. Reducing sedentary behavior: The relationship between park area and the physical activity of youth. Psychol. Sci. 17(8):654-659.

Ewing, R., and R. Cervero. 2001. Travel and the built environment: A synthesis. Transportation Research Record 1780:87-114.

Ewing, R., T. Schmid, R. Killingsworth, A. Zlot, and S. Raudenbush. 2003. Relationship between urban sprawl and physical activity, obesity, and morbidity. Am. J. Health Promot. 18(1):47-57.

Ferrie, J.E., M.J. Shipley, M.G. Marmot, S. Stansfeld, and G.D. Smith. 1998. The health effects of major organizational change and job insecurity. Soc. Sci. Med. 46(2):243-254.

Ferrie, J.E., H. Westerlund, M. Virtanen, J. Vahtera, and M. Kivimäki. 2008. Flexible labor markets and employee health. Scand. J. Work Environ. Health 34(6):98-110.

Fields, S. 2004. The fat of the land: Do agricultural subsidies foster poor health? Environ. Health Perspect. 112(14):A820-A823.

Fitzhugh, E.C., D.R. Bassett, Jr., and M.F. Evans. 2010. Urban trails and physical activity: A natural experiment. Am. J. Prev. Med. 39(3):259-262.

Fleishman, J.A. 2005. Demographic and Clinical Variations in Health Status. MEPS Methodology Report No. 15. Agency for Healthcare Research and Quality, Rockville, MD [online]. Available: http://www.meps.ahrq.gov/mepsweb/data_files/publications/mr15/mr15.pdf [accessed May 11, 2011].

Frank, J. and K. Cullen. 2006. Preventing injury, illness and disability at work. Scand. J. Work Environ. Health. 32(2):160-167.

Frank, L.D., and P. Engelke. 2005. Multiple impacts of the built environment on public health: Walkable places and the exposure to air pollution. Int. Reg. Sci. Rev. 28(2):193-216.

Frank, L., and S. Kavage. 2009. A national plan for physical activity: The enabling role of the built environment. J. Phys. Act Health. 6(suppl. 2):S186-S195.

Frank, L.D., B. Stone, and W. Bachman. 2000. Linking land use with household vehicle emissions in the central Puget sound: Methodological framework and findings. Transport. Res. D-Tr. E. 5(3):173-196.

Frank, L.D., M.A. Andresen, and T.L. Schmid. 2004. Obesity relationships with community design, physical activity, and time spent in cars. Am. J. Prev. Med. 27(2):87-96.

Frank, L.D., T. Schmid, J.F. Sallis, J. Chapman, and B. Saelens. 2005. Linking objectively measured physical activity data with objectively measured urban form: Findings from SMARTRAQ. Am. J. Prev. Med. 28(suppl. 2):117-125.

Frank, L.D., J.F. Sallis, T. Conway, J. Chapman, B. Saelens, and W. Bachman. 2006. Multiple pathways from land use to health: Association between neighborhood walkability and active transportation, body mass index, and air quality. J. Am. Plann. Assoc. 72(1):75-87.

Frank, L.D., M. Bradley, S. Kavage, J. Chapman, and T.K. Lawton. 2007. Urban form, travel time, and cost relationships with tour complexity and mode choice. Transportation 35(1):37-54.

Franzini, L., M.N. Elliott, P. Cuccaro, M. Schuster, M.J. Gilliland, J.A. Grunbaum, F. Franklin, and S.R. Tortolero. 2009. Influences of physical and social neighborhood environments on child physical activity and obesity. Am. J. Public Health 99(2):271-278.

Friedman, M.S., K.E. Powell, L. Hutwagner, L.M. Graham, G. Teague. 2001. Impact of changes in transportation and commuting behaviors during the 1996 Summer Olympic Games in Atlanta on air quality and childhood asthma. JAMA 285(7):897-905.

Frohlich, K.L., and L. Potvin. 2008. The inequality paradox: The population approach and vulnerable populations. Am. J. Public Health 98(2):216-221.

Gauderman, W.J., E. Avol, F. Lurmann, N. Kuenzli, F. Gilliland, J. Peters, and R. McConnell. 2005. Childhood asthma and exposure to traffic and nitrogen dioxide. Epidemiology 16(6):737-743.

Gillen, M., I.H. Yen, L. Trupin, L. Swig, R. Rugulies, K. Mullen, A. Font, D. Burian, G. Ryan, I. Janowitz, P.A. Quinlan, J. Frank, and P. Blanc. 2007. The association of socioeconomic status and psychosocial and physical workplace factors with musculoskeletal injury in hospital workers. Am. J. Ind. Med. 50(4):245-260.

Gordon-Larsen, P., M.C. Nelson, and K. Beam. 2005. Associations among active transportation, physical activity, and weight status in young adults. Obes. Res. 13(5):868-875.

Green, R.S., S. Smorodinsky, J.J. Kim, R. McLaughlin, and B. Ostro. 2004. Proximity of California public schools to busy roads. Environ Health Perspect. 112(1):61-66.

Gunier, R.B., A. Hertz, J. Von Behren, and P. Reynolds. 2003. Traffic density in California: Socioeconomic and ethnic differences among potentially exposed children. J. Expo. Anal. Environ. Epidemiol. 13(3):240-246.

Hammitt, J.K. 2007. Valuing changes in mortality risk: Lives saved vs. life years saved. Rev. Environ. Econ. Policy 1(2):228-240.

Hannon, C., A. Cradock, S.L. Gortmaker, J. Wiecha, A. El Ayadi, L. Keefe, and A. Harris. 2006. Play across Boston: A community initiative to reduce disparities in access to after-school physical activity programs for inner-city youths. Prev. Chronic Dis. 3(3):A100.

Hawkes, C. 2007. Promoting healthy diets and tackling obesity and diet-related chronic diseases: What are the agricultural policy levers? Food Nutr. Bull. 28(suppl. 2):S312-S322.

Houston, D., J. Wu, P. Ong, and A. Winer. 2004. Structural disparities of urban traffic in southern California: Implications for vehicle-related air pollution exposure in minority and high-poverty neighborhoods. J. Urban Aff. 26(5):565-592.

Jacobsen, J.O., N.W. Hengartner, and T.A. Louis. 2004. Inequity Measures for Evaluation of Environmental Justice: A Case Study of Close Proximity to Highways in NYC. Paper 29. Johns Hopkins University, Department of Biostatics Working Papers [online]. Available: http://www.bepress.com/cgi/viewcontent.cgi?article=1029&context=jhubiostat [accessed May 11, 2011].

Jerrett, M., K. Shankardass, K. Berhane, W.J. Gauderman, N. Künzli, E. Avol, F. Gilliland, F. Lurmann, J.N. Molitor, J.T. Molitor, D.C. Thomas, J. Peters, and R. McConnell. 2008. Traffic-related air pollution and asthma onset in children: A prospective cohort study with individual exposure measurement. Environ. Health Perspect. 116(10):1433-1438.

Kawachi, I., N.E. Adler, and W.H. Dow. 2010. Money, schooling, and health: Mechanisms and causal evidence. Ann. NY Acad. Sci. 1186:56-68.

King, H.C., D.L. Hoyert, J.Q. Xu, and S.L. Murphy. 2008. Deaths: Final Data for 2005. National Vital Statistics Reports 56(10). Hyattsville, MD: National Center for Health Statistics.

Krieger, N. 2010. Workers are people too: Societal aspects of occupational disparities— an ecosocial perspective. Am. J. Ind. Med. 53(2):104-115.

Larson, N.I., M.T. Story, and M.C. Nelson. 2009. Neighborhood environments: Disparities in access to healthy foods in the U.S. Am. J. Prev. Med. 36(1):74-81.

Ledikwe, J.H., H.M. Blanck, L. Kettel Khan, M.K. Serdula, J.D. Seymour, B.C. Tohill, and B.J. Rolls. 2006. Dietary energy density is associated with energy intake and weight status in US adults. Am. J. Clin. Nutr. 83(6):1362-1368.

Li, F., P.A. Harmer, B.J. Cardinal, M. Bosworth, A. Acock, D. Johnson-Shelton, and J.M. Moore. 2008. Built environment, adiposity, and physical activity in adults aged 50-75. Am. J. Prev. Med. 35(1):38-46.

Link, B.G., and J. Phelan. 1995. Social conditions as fundamental causes of disease. J. Health Soc. Behav. (Spec No):80-94.

Lleras-Muney, A. 2005. The relationship between education and adult mortality in the United States. Rev. Econ. Stud. 72(1):189-221.

Lopez, R. 2004. Urban sprawl and risk for being overweight or obese. Am. J. Public Health 94(9):1574-1579.

Lynch, J.W., G.A. Kaplan, R.D. Cohen, J. Tuomilehto, and J.T. Salonen. 1996. Do cardiovascular risk factors explain the relation between socioeconomic status, risk of all-cause mortality, cardiovascular mortality, and acute myocardial infarction? Am. J. Epidemiol. 144(10):934-942.

Mackenbach, J.P., W.J. Meerding, and A.E. Kunst. 2007. Economic Implications of Socioeconomic Inequalities in Health in the European Union. Erasmus MC, Department of Public Health, Rotterdam, The Netherlands [online]. Available: http://survey.erasmusmc.nl/intern/pwp/upload/24/Final%20report%20Macroeconomics%20July%202007-1.pdf [accessed Aug. 9, 2011].

Mann, J.K., J.R. Balmes, T.A. Bruckner, K.M. Mortimer, H.G. Margolis, B. Pratt, S.K. Hammond, F.W. Lurmann, and I.B. Tager. 2010. Short-term effects of air pollution on wheeze in asthmatic children in Fresno, California. Environ. Health Perspect. 118(10):1497-1502.

Mar, T.F., J.Q. Koenig, and J. Primomo. 2010. Associations between asthma emergency visits and particulate matter sources, including diesel emissions from stationary generators in Tacoma, Washington. Inhal. Toxicol. 22(6):445-448.

Marmot, M., and T. Theorell. 1988. Social class and cardiovascular disease: The contribution of work. Int. J. Health Serv. 18(4):659-674.

Marmot, M., M. Shipley, E. Brunner, and H. Hemingway. 2001. Relative contribution of early life and adult socioeconomic factors to adult morbidity in the Whitehall II study. J. Epidemiol. Community Health 55(5):301-307.

Mendoza, J.A., A. Drewnowski, and D.A. Christakis. 2007. Dietary energy density is associated with obesity and the metabolic syndrome in U.S. adults. Diabetes Care 30(4):974-979.

Meyer, J.D., N. Warren, and S. Reisine. 2010. Racial and ethnic disparities in low birth weight delivery associated with maternal occupational characteristics. Am. J. Ind. Med. 53(2):153-162.

NAGC (National Sustainable Agriculture Coalition). 2010. Healthy Food Financing Initiative Introduced in House, Senate, December 1, 2010. National Sustainable Agriculture Coalition [online]. Available: http://sustainableagriculture.net/blog/hffi-bill-introduced/ [accessed Jan. 11, 2011].

NHTSA (National Highway Traffic Safety Administration). 2010. Fatality Analysis Reporting System Encyclopedia. U.S. Department of Transportation, National Highway Traffic Safety Administration [online]. Available: http://www-fars.nhtsa.dot.gov/Main/index.aspx [accessed Jan. 11, 2011].

NRC (National Research Council). 2009. Science and Decisions: Advancing Risk Assessment. Washington, DC: National Academies Press.

NRC (National Research Council). 2010. Hidden Costs of Energy: Unpriced Consequences of Energy Production and Use. Washington, DC: National Academies Press.

Ogden, C.L., M.D. Carroll, M.A. McDowell, and K.M. Flegal. 2007. Obesity Among Adults in the United States—No Statistically Significant Change Since 2003-2004. NCHS Data Brief no 1. Hyattsville, MD: National Center for Health Statistics [online]. Available: http://www.cdc.gov/nchs/data/databriefs/db01.pdf [accessed May 12, 2011].

Ogden, C.L., M.D. Carroll, and K.M. Flegal. 2008. High body mass index for age among U.S. children and adolescents, 2003-2006. JAMA 299(20):2401-2405.

Ostry, A.S., S. Radi, A.M. Louie, and A.D. LaMontagne. 2006. Psychosocial and other working conditions in relation to body mass index in a representative sample of Australian workers. BMC Public Health 6:53-60.

Park, S.K., A.H. Auchincloss, M.S. O'Neill, R. Prineas, J.C. Correa, J. Keeler, R.G. Barr, J.D. Kaufman, and A.V. Diez-Roux. 2010. Particulate air pollution, metabolic syndrome and heart rate variability: The multi-ethnic study of atherosclerosis (MESA). Environ. Health Perspect. 118(10):1406-1411.

Pohanka, M., and S. Fitzgerald. 2004. Urban sprawl and you: How sprawl adversely affects worker health. AAOHN J. 52(6):242-246.

Ponce, N.A., K.J. Hoggatt, M. Wilhelm, and B. Ritz. 2005. Preterm birth: The interaction of traffic-related air pollution with economic hardship in Los Angeles neighborhoods. Am. J. Epidemiol. 162(2):140-148.

Pope III, C.A., and D.W. Dockery. 2006. Health effects of fine particulate air pollution: Lines that connect. J. Air Waste Manage. Assoc. 56(6):709-742.

Ritz, B., M. Wilhelm, K.J. Hoggatt, and J.K. Ghosh. 2007. Ambient air pollution and preterm birth in the Environment and Pregnancy Outcomes Study at the University of California, Los Angeles. Am. J. Epidemiol. 166(9):1045-1052.

Roberts, E.M. 1997. Neighborhood social environments and the distribution of low birthweight in Chicago. Am. J. Public Health 87(4):597–603.

Rosenstock, L., M.R. Cullen, C.A. Brodkin, and C.A. Redlich, eds. 2005. Textbook of Clinical Occupational and Environmental Medicine. Philadelphia: W.B. Saunders.

RWJF (Robert Wood Johnson Foundation). 2009. Beyond Health Care: New Directions to a Healthier America. Robert Wood Johnson Foundation Commission to Build a

Healthier America [online]. Available: http://www.rwjf.org/pr/product.jsp?id=41008 [accessed Jan. 10, 2011].

Saelens, B.E., J.F. Sallis, J.B. Black, and D. Chen. 2003. Neighborhood-based differences in physical activity: An environment scale evaluation. Am. J. Public Health 93(9):1552-1558.

Salam, M.T., J. Millstein, Y.F. Li, F.W. Lurmann, H.G. Margolis, and F.D. Gilliland. 2005. Birth outcomes and prenatal exposure to ozone, carbon monoxide, and particulate matter: Results from the Children's Health Study. Environ. Health Perspect. 113(11):1638-1644.

Schulte, P.A., G.R. Wagner, A. Ostry, L.A. Blanciforti, R.G. Cutlip, K.M. Krajnak, M. Luster, A.E. Munson, J.P. O'Callaghan, C.G. Parks, P.P. Simeonova, and D.B. Miller. 2007. Work, obesity, and occupational safety and health. Am. J. Public Health 97(3):428-436.

Schulz, A., and M. Northridge. 2004. Social determinants of health: Implications for environmental health promotion. Health Educ. Behav. 31(4):455-471.

Siegrist, J. 1996. Adverse health effects of high-effort/low-reward conditions. J. Occup. Health Psychol. 1(1):27-41.

Slama, R., V. Morgenstern, J. Cyrys, A. Zutavern, O. Herbarth, H.E. Wichmann, J. Heinrich, and LISA Study Group. 2007. Traffic-related atmospheric pollutant levels during pregnancy and offspring's term birth weight: A study relying on a land-use regression exposure model. Environ. Health Perspect. 115(9):1283-1292.

Sturm, R. 2008. Disparities in the food environment surrounding U.S. middle and high schools. Public Health 122(7):681-690.

Tillotson, J.E. 2004. America's obesity: Conflicting public policies, industrial economic development, and unintended human consequences. Ann. Rev. Nutr. 24:617-643.

TRB (Transportation Research Board). 2005. Does the Built Environment Influence Physical Activity? Examining the Evidence. Transportation Research Board Special Report 282. Washington, DC: Transportation Research Board.

Truong, K., M. Fernandes, R. An, V. Shier, and R. Sturm. 2010. Measuring the physical food environment and its relationship with obesity: Evidence from California. Public Health 124(2):115-118.

UCBHIG (University of California Berkeley Health Impact Group). 2007. Oak to Ninth Avenue HIA. University of California Berkeley Health Impact Group, June 2007 [online]. Available: http://sites.google.com/site/ucbhia/projects-and-research [accessed June 2, 2010].

Wang, J., R. Luben, K.T. Khaw, S. Bingham, N.J. Wareham, and N.G. Forouhi. 2008. Dietary energy density predicts the risk of incident type 2 diabetes: The EPIC-Norfolk study. Diabetes Care 31(11):2120-2125.

WHO (World Health Organization). 2001. Macroeconomics and Health: Investing in Health for Economic Development. Report of the Commission on Microeconomics and Health. Geneva: World Health Organization [online]. Available: http://www.paho.org/english/hdp/hdd/sachs.pdf [accessed May 13, 2011].

WHO (World Health Organization). 2002. The World Health Report 2002: Reducing Risks, Promoting Healthy Life. Geneva: World Health Organization [online]. Available: http://www.who.int/whr/2002/en/index.html [accessed Jan. 10, 2011].

WHO (World Health Organization). 2010. World Health Statistics: 2010. Geneva: World Health Organization [online]. Available: http://www.who.int/whosis/whostat/en/ [accessed Jan. 10, 2011].

Wier, M., J. Weintraub, E.H. Humphreys, E. Seto, and R. Bhatia. 2009. An area-level model of vehicle-pedestrian injury collisions with implications for land use and transportation planning. Accid. Anal. Prev. 41(1):137-145.

Williams, D.R., S.A. Mohammed, J. Leavell, and C. Collins. 2010. Race, socioeconomic status, and health: Complexities, ongoing challenges, and research opportunities. Ann. NY Acad. Sci. 1186:69-101.

Woodruff, T.J., L.A. Darrow, and J.D. Parker. 2008. Air pollution and postneonatal infant mortality in the United States, 1999-2002. Environ. Health Perspect. 116(1):110-115.

Wu, S.Y., and A. Green. 2000. Projection of Chronic Illness Prevalence and Cost Inflation. Santa Monica, CA: RAND.

Wu, Y.C., and S.A. Batterman. 2006. Proximity of schools in Detroit, Michigan to automobile and truck traffic. J. Expo. Sci. Environ. Epidemiol. 16(5):457-470.

3

Elements of a Health Impact Assessment

Chapter 2 established the rationale for examining the potential effects of decisions on health and health disparities and highlighted health impact assessment (HIA) as a potential tool for assessing the health implications of various decisions. This chapter describes the types, structure, and content of HIAs and summarizes the HIA process, methodologic approaches, and variations in practice. It is informed by a review of U.S. and international HIA literature and guidelines (see Appendixes A and E) and by the experience of committee members and others who provided input during the committee process. On the basis of its review, the committee synthesized the information from guidance, practice, and literature to propose criteria that define an HIA and draw several conclusions regarding HIA practice. As discussed in this chapter, HIAs have been used for a wide variety of applications and at all levels of government (local, state, tribal, and federal) and have been conducted with varied resources over different schedules. The committee does not intend that the definition and criteria proposed in this chapter be considered rigid requirements but rather that they reflect an ideal of practice, deviation from which may occur but should be based on clear and well-articulated needs and rationale.

Before discussing the various elements of HIA, it is important to understand the context in which HIA is undertaken in the United States. As described in Appendix A, there are few laws in the United States that specifically require HIA, although many—such as the National Environmental Policy Act (NEPA)—require a consideration of health that can be accomplished through HIA. Most HIAs in the United States are therefore undertaken outside the formal decision-making process by organizations (such as nonprofit community-based groups), universities, or health departments that do not have decision-making authority over the proposals being addressed. Although less common to date, HIAs are also sometimes conducted by a decision-making agency, such as a metropolitan planning organization or a federal agency complying with NEPA. The decision to initiate an HIA is often made ad hoc when public-health advocates recognize that the proposal may have important health implications that would not otherwise be recognized or addressed. There are often not clear lines of authority between the team conducting the HIA and the decision-maker. The health effects that are included, the data sources and methods that are used, and the recommendations that are made are therefore determined by the HIA practitioners rather than according to a legal or regulatory standard (Wernham 2011). Thus, the assessment phase is separated from the management phase, as recommended elsewhere (NRC 1983). The fact, however, that the team conducting the HIA is aware of the decision context allows the assessment to be decision-relevant.

CATEGORIES OF HEALTH IMPACT ASSESSMENT

Scholars point to a remarkable consistency in the basic elements that are generally included in descriptions of HIA (Mindell et al. 2008). In practice, however, there is some inconsistency in how HIAs are conducted—for example, how stakeholders are engaged and how data are collected and analyzed—and in the structure and content of the final work products of an HIA. The diversity of practice owes partly to the fact that HIAs are undertaken for a wide array of policy-making that spans many sectors, levels of government, types of proposal (policies, plans, programs, and projects), and degrees of complexity. The variability in the practice has evolved in the absence of widely accepted practice standards or formal regulatory or procedural requirements for HIA outside NEPA and related state laws (see Appendix A). However, it appears to be increasingly accepted that HIA is carried out to inform the decision rather than to evaluate the impacts after the decision is made, and there is general agreement on the procedural steps of HIA (Harris-Roxas and Harris 2011).

HIA practice is often defined in terms of several categories. According to effort, complexity, and duration, HIAs are often described as rapid, intermediate, or comprehensive. Rapid HIAs may be completed in a short time (weeks to months), are often focused on smaller and less complex proposals, and generally involve primarily literature review and descriptive or qualitative analysis. The phrase *desktop HIA* has also been used to refer to a rapid HIA that entails little or no public engagement. Another variation, rapid-appraisal HIA, has been described and in some texts includes explicit public engagement through an initial half-day workshop for stakeholders (Parry and Stevens 2001; Mindell et al. 2003; ICMM 2010). Intermediate HIAs require more time and resources and involve more complex pathways, more stakeholder engagement, and a more detailed analysis but include little collection of new data. Comprehensive HIAs are most commonly differentiated from rapid and intermediate HIAs by the scope of potential impacts and the need for collection of new primary data. They can take longer than a year to complete.

HIAs are also differentiated according to whether they are integrated into an environmental impact assessment or done independently. Another categorization is based on the breadth of the HIA and distinguishes HIAs that have a tight focus—such as ones that use a narrow definition of health and emphasize quantification—from HIAs that have a broader, holistic focus shaped by the social determinants of health (Kemm 2001). Others have proposed categorizing HIAs as participatory (emphasizing shared governance, public participation, and a focus on socioeconomic and environmental determinants), quantitative or analytic (concentrating on the methods and rigor of the analysis), or procedural (drawing on elements of the other two approaches but emphasizing the procedural steps required and often undertaken within a specified administrative or regulatory context) (Cole and Fielding 2007).

In practice, the categories are rarely used consistently, and a single HIA often encompasses a blend of various approaches to stakeholder engagement and participation, analytic methods, and interactions with the formal decision-making process. For example, desktop HIAs may consider indirect stakeholder input through review of public comments submitted outside the HIA process, comprehensive HIAs may have relatively little stakeholder engagement, and rapid-appraisal HIAs of smaller-scale proposals may involve collection of some new data to inform the analysis. The various categories of HIAs, although useful for describing distinct themes in the field, do not necessarily represent consistently distinct strains of practice. Instead, it appears that the specific methods and approaches used in a single HIA often evolve within the basic framework described above and develop as a pragmatic response to context. Influences on practice include the timeline, resources and skills available to the HIA team, the factors being considered and the data available for analysis, and the legal and regulatory context of the decision-making process. That description is consistent with the earlier characterizations of HIA as a combination of procedures, methods, and tools (WHO 1999; Quigley et al. 2006).

The committee notes that the diversity of approaches and decision contexts imposes challenges for determining the resources required for conducting an HIA. For example, although rapid HIAs are small-scale, low-cost investigations, comprehensive HIAs that require new primary data collection can take longer than a year to complete and require substantially more resources. Information on costs of HIAs would be valuable in determining whether an HIA can be undertaken with the resources available and could inform the screening process as described below. However, the committee notes that no published studies in the United States have attempted to quantify the costs of undertaking an HIA across a variety of settings. Such information would be useful for informing future implementation.

DEFINITION OF HEALTH IMPACT ASSESSMENT

The committee proposes on the basis of its review the following adaptation of the current working definition of the International Association of Impact Assessment (Quigley et al. 2006) as a technical definition of HIA:

HIA is a systematic process that uses an array of data sources and analytic methods and considers input from stakeholders to determine the potential effects of a proposed policy, plan, program, or project on the health of a population and the distribution of the effects within the population. HIA provides recommendations on monitoring and managing those effects.

That definition reflects the committee's finding that the involvement of stakeholders—although the approaches used vary from little or no involvement to robust engagement and participation at every step—has consistently been described as a core element of HIA practice and should be considered essential to it. Although rapid or desktop HIAs may not involve stakeholders or consider their input, this often (although not uniformly) reflects a pragmatic response to limitations, such as the timeframe for the decision or resources available to the HIA team, rather than an optimal practice. The definition also notes that recommendations should incorporate monitoring, which is essential for effective continuing management as a decision is implemented.

WHO CONDUCTS HEALTH IMPACT ASSESSMENTS?

HIAs can be conducted by a variety of agencies, organizations, or individuals. A decision-making body—such as a department of planning or transportation—can conduct an HIA to inform its own decision. It is also common for local, state, or tribal health departments to undertake an HIA to inform another agency's decision-making. University researchers have conducted HIAs, and community-based organizations have conducted HIAs with technical assistance from public-health experts to inform officials who are deliberating on a legislative or administrative proposal. HIAs are also done by private consultants who are hired by a project proponent or decision-maker or by private-industry stakeholders.

Because the assessment of health effects depends on an in-depth understanding of changes that may affect health—such as changes in traffic flow, roadway design, air quality, or community revenue sources—HIAs are inherently multidisciplinary; public-health experts may lead the effort but must draw on resources and expertise from other disciplines. Thus, HIA teams may include not only health experts but professionals in other related disciplines, such as air or water quality or traffic modeling. As discussed in greater depth in the section on scoping, it is common to convene advisory or steering committees, which can include both technical and policy experts and representatives from stakeholder groups that have an interest in the decision outcome.

The training and credentials of HIA practitioners are variable, and there is no universally accepted standard for a level of training necessary to lead an HIA. In the United States, HIAs have commonly been undertaken by people who have an MPH or equivalent degree and have attended a brief (2- to 5-day) training session. In this report, *HIA practitioner* refers to the person (or people) involved in conducting an HIA.

PROCESS FOR HEALTH IMPACT ASSESSMENT

The tasks or elements that are described as part of an HIA are fairly consistent in the peer-reviewed literature and guides reviewed by the committee. The grouping of the elements in discrete stages or steps of an HIA is less consistent; some guides list as few as five steps, and others describe as many as nine (Quigley et al. 2006; Bhatia 2010; ICMM 2010). The committee selected a six-step framework as a clear way to organize and describe the critical elements of an HIA. The steps can be described as follows:

(1) *Screening* determines whether a proposal is likely to have health effects and whether the HIA will provide information useful to the stakeholders and decision-makers.

(2) *Scoping* establishes the scope of health effects that will be included in the HIA, the populations affected, the HIA team, sources of data, methods to be used, and alternatives to be considered.

(3) *Assessment* involves a two-step process that first describes the baseline health status of the affected population and then assesses potential impacts.

(4) *Recommendations* suggest design alternatives that could be implemented to improve health or actions that could be taken to manage the health effects, if any, that are identified.

(5) *Reporting* documents and presents the findings and recommendations to stakeholders and decision-makers.

(6) *Monitoring and evaluation* are variably grouped and described. Monitoring can include monitoring of the adoption and implementation of HIA recommendations or monitoring of changes in health or health determinants. Evaluation can address the process, impact, or outcomes of an HIA.

The following sections provide an overview of the process of conducting an HIA. For each step, the committee describes the basic purpose, objectives, and practice elements; summarizes the main outputs; and presents conclusions regarding pertinent issues raised. Major issues and challenges for HIA development and practice are considered in Chapter 4. The reader will notice that some of the committee's descriptions and characterizations overlap with those of other guides; the similarities highlight the consistencies in the field.

Screening

Screening establishes the need for and value of conducting an HIA. Because HIAs can address decisions that range from small, localized programs or projects to national policies, screening ensures that HIA is used judiciously and when it is most likely to be valuable. Given the volume and breadth of decisions at the local, state, tribal, and federal levels that can potentially affect health in some way, one of the challenges for HIA practice is to determine which proposals to screen. In the absence of mandates or formal procedures, topics for screening are often chosen on the basis of the interests of a group wishing to use HIA as opposed to a structured, strategic selection process.[1]

More structured approaches have also been used. In some cases, collaboration between a health department and other agencies has resulted in the identification of appropriate proposals for screening. In other cases, all proposals in selected agencies or sectors have been screened by local governments (SFCC 1998; Lester et al. 1999; Roscam Abbing 2004). For example, the San Francisco Department of Public Health routinely screens major projects and plans to ensure adequate analysis and mitigation of environmental health impacts. In Alaska, all large natural-resources development proposals are now screened for the need for HIA in a new program begun by the state health department.

Screening involves making an initial rapid judgment of whether an HIA is likely to be feasible and valuable. The central considerations include whether the proposal in question might cause important changes in health, whether health is already a major focus of the decision-making process, whether the legal framework provides an opportunity for health to be factored into the decision, and whether data, staff, resources, and time are adequate to complete a successful HIA in time to provide useful input into the decision-making process (that is, can information be provided within the timeline for the decision). Another consideration is whether the proposal is likely to place a disproportionate burden of risk on vulnerable populations in the affected community; screening proposals on this basis helps to ensure that the HIA addresses the risk factors that underlie observed disparities in the rates of illness among various populations.

A variety of screening tools and algorithms are commonly used (Cole et al. 2005; PHAC 2005; Harris et al. 2007; Bhatia 2010). Some use pertinent screening questions, such as the ones noted, and apply a sequential yes-no query to each (Cole et al. 2005). Some provide a checklist of factors to consider and often focus on health determinants that might be affected by the proposal. Some decisions to conduct HIA may depend on a specific statutory requirement or mandated procedure. For example, in the context of NEPA, the lead federal agency must consider "the degree to which the proposed action affects public health or safety" to determine whether a proposal is likely to have "significant" effects and therefore require an environmental impact statement (40 CFR

---

[1] Under NEPA, a federal agency must determine whether a federal environmental decision is likely to have significant effects, and if so, the level of analysis required (40 C.F.R. Section 1508.27). Because the degree to which the proposed action affects public health or safety is one factor considered, this process could be considered the equivalent of the screening step of an HIA. In practice, however, explicit consideration of health has been rare (Steinemann 2000; Cole et al. 2004; Bhatia and Wernham 2008).

1508.27). Ultimately, regardless of the specific tool used, the decision to conduct HIA in most cases relies on the practitioner's or decision-maker's judgment regarding the likelihood of impacts, the time and resources available, and the likelihood that the information produced by the HIA will be a valuable aid to decision-making.

Because any actions taken on the basis of HIA recommendations need to be implemented within a specific legal and policy context, screening needs to establish a clear description of the decision-making process and context. It should also identify the points at which there is an opportunity for information from the HIA to influence decisions. Mapping out the timeline for the decision-making process can be helpful, and for large and complex programs and projects, identifying the agencies involved and their jurisdictions is important. Such programs and projects involve many agencies and entities that have authority over some aspect of planning and implementation. For example, the planning of the Atlanta Beltline, as described later in this chapter, involved the regional planning commission, local legislative bodies, state and federal environmental regulators, and private developers. It is also useful to assess the political context of the proposal to be assessed and consider, for example, the major political drivers of the proposal, the arguments made by political supporters and those opposed to the proposal, and any economic or technical constraints that limit the alternatives that can be considered.

Public concerns are a common trigger for a decision to screen, and the degree of concern or controversy about a proposal may be one of the factors weighed in the decision to undertake an HIA. For example, the Massachusetts Department of Public Health responded to citizen concerns regarding a proposed power plant by considering whether HIA would be an appropriate way to address them (McAuliffe 2009). The committee notes that public involvement is important in screening; information provided by stakeholders may provide insight into the potential effects of a proposal under consideration that contribute to the final determination of whether an HIA is warranted and likely to be useful.

Screening is often not well documented, and it is often not clear from an HIA report what factors were considered in making the decision to do an HIA. Moreover, because there is generally no written record of HIAs that stop at screening, still less is known about the reasons that have led to decisions not to proceed with HIA. Box 3-1 provides an example of how screening on a proposal for a residential housing program was conducted. It includes the information that was taken into account and the final output of the screening process, which was a decision on whether to commission and proceed with an HIA.

Outputs of Screening

Screening should result in a simple statement that includes the following:

- A description of the proposed policy, program, plan, or project that will be the focus of the HIA, including the timeline for the decision and intervention points at which HIA information will be used.
- A statement of why the proposal was selected for screening.
- A preliminary opinion regarding the potential importance of the proposal for health.
- The expected resource requirements of the HIA and the ability of the HIA team to meet them.
- A description of the political and policy context of the decision and an analysis of the opportunities to influence decision-making or otherwise make health-oriented changes.

BOX 3-1 Screening: HIA of a Residential Housing Program

The Crossings is a proposed housing development in Los Angeles that will provide 450 units in a newly rezoned residential area that needs affordable housing. A local community-based organization worked with a housing developer on the proposal and site plan. They expressed "interest in developing The Crossings in a way that will address local community needs for affordable housing and for other community assets that are safe, healthy, and supportive" (p. Intro-1).

In 2009, an HIA was conducted to ensure that health impacts were considered in the design and development of The Crossings and in the broader policies that affected redevelopment in the area. The HIA report describes the screening process but does not provide great detail about it.

The HIA notes that the area within which The Crossings is proposed to be built has the following characteristics:

- A growing population of families that have children.
- Dilapidated housing conditions.
- Prevalence of overcrowding.
- A lack of access to needed goods and services.

The HIA notes that the residential area is inhabited by a vulnerable population, that the built environment is of low quality, that the development will potentially have important health implications for residents in the local and surrounding communities, and that there is a strong commitment shown by the community and the developer to integrate health considerations into the planning process. It was concluded during the screening phase that an HIA would add value to project outcomes. An HIA would identify health assets, health liabilities, and health-promoting mitigations related to the proposed development project. The facts that resources were available and that timelines were appropriate were also relevant to the decision to conduct an HIA.

Source: Adapted from Heller et al. 2009.

- A screening recommendation—for example, no further action required; no HIA, but health advice and input to be offered in an alternative way; or proceed with HIA.

Committee Conclusions Regarding Screening

Screening is essential for high-quality HIA. Poorly selected proposals may result in HIAs that add little new information and consume considerable time and resources of the HIA team to complete and of recipients to review. HIA should not be assumed to be the best approach to every health-policy question but should instead be seen as part of a spectrum of public-health and policy-oriented approaches, some of which will be more appropriate than others, depending on the specific application. Although the reasons and objectives for HIA are often not articulated at the outset of screening, establishing well-defined objectives will focus the screening process on determining whether HIA is likely to be an effective approach for achieving them.

Any approach to determining which proposals will be screened should demonstrate a consistent rationale; should document the rationale in the HIA report; and should take account of public input. Screening should also consider whether a proposal conforms with applicable standards, policies, or laws relevant to health inasmuch as there is a wide variety of them that bear directly or indirectly on health. For example, U.S. priorities for improving public health are expressed in the Healthy People 2020 Program of the U.S. Department of Health and Human Services (DHHS 2010). Some laws—such as NEPA, state environmental-policy acts, and various local zoning ordinances—may establish protection of health as a requirement or priority. The programs and policies, however, may not provide any guidance on how health should be considered (see, for example, Pub. L 91-190, 42 U.S.C. 4321-4347 [1970]; EC 2001). Furthermore, some policies may focus on determinants of health—for example, economic development, transportation, or housing—rather than explicitly mentioning health. In each case, it is important to determine how the standards, policies, programs, and laws bear on how health is factored into a proposal.

The committee concludes that the following are the most important factors to consider in determining whether to do an HIA:

- The potential for substantial adverse or beneficial health effects and the potential to make changes in the proposal that could result in an improved health risk-benefit profile.
- The potential for HIA-based information to alter a decision or help a decision-maker discriminate among decision options.
- The potential for irreversible or catastrophic effects (including effects of low likelihood).
- The potential for health effects to place a disproportionate burden on or substantially benefit vulnerable populations.

- Public concern or controversy regarding health effects of the proposed decision.
- The opportunity to bring health information into a decision-making process that may otherwise not include this information.
- The potential for the HIA to be completed in the time allotted and with the resources available.

Ultimately, the HIA report should provide a rational and consistent explanation of how proposals are selected for screening. That explanation is particularly important when public funds are to be used for an HIA because the public may want to understand the basis for allocating sparse public resources. Given the breadth of decisions that are likely to warrant consideration, the approach taken will vary on the basis of who is initiating the HIA, the capacity and authority of the agency or entity undertaking it, and the objectives for contemplating an HIA.

Scoping

Scoping establishes the boundaries of the HIA and identifies the health effects to be evaluated, the populations affected, the HIA team, sources of data, methods to be used, and any alternatives to be assessed. Well-executed scoping saves time, work, and resources in the later stages of the HIA (Harris et al. 2007). The choice of what to evaluate will reflect the specific social, political, and policy context of the decision; the needs, interests, and questions of stakeholders and decision-makers; and the health status of the affected population.

Potential Health Effects

Determining the potential health effects to include in the HIA and proposing hypothetical causal pathways are the central tasks of scoping. Scoping considers input from many sources, including preliminary literature searches, public input, and professional or expert opinion in fields relevant to the proposal. Because it will often not be practical or possible to address all direct and indirect health effects that appear theoretically possible, it is important to select issues carefully.[2] Setting priorities considers pathways that appear most important from a public-health perspective and considers issues that have been raised prominently by stakeholders. Questions that are important from a public-health perspective might include the severity of the health effect, the size and likelihood of the effect, and the potential of the effect to exacerbate health disparities. In practice, some HIAs have focused on a specific health end point, such as obesity, or

---

[2]Identifying high-priority issues has been addressed in numerous contexts outside HIA, including human-health and ecologic risk assessment (see, for example, EPA 1989, 1992; NRC 1996, 2009).

health concerns related to a single impact of the proposal, such as the health effects of air pollutants, most likely without using a systematic approach that considered and eliminated other impacts (see, for example, Kuo et al. 2009; Castro et al. 2010).

Iteration during scoping and between scoping and assessment often results in additional changes in the final list of issues included in the HIA. During scoping, the HIA team may produce an initial list, refine it on the basis of stakeholder input, and then make it final through research and analysis in the assessment phase. In other cases, the initial scope is generated by stakeholders and then refined through research and input from advisory or steering committees.

Several approaches for scoping are available. One approach uses a logic framework that maps out the causal pathways by which health effects might occur (see Figure 3-1). In general, this approach describes effects directly related to the proposal (such as changes in air emissions) and traces them to health determinants (such as air quality) and finally to health outcomes (such as asthma). The first step in the framework is typically a determinant of health, such as air pollution, traffic, employment, or noise. Logic frameworks can be used as part of stakeholder engagement to develop a shared understanding of how a project will develop and the outcomes that can be expected (Cave and Curtis 2001a,b; Cave et al. 2001). Another method of scoping is to develop a table that facilitates a systematic and rapid appraisal of all the potential ways in which a proposal might affect health (see Table 3-1). In this approach, the aspects of a proposal that may affect health are listed and considered in major categories of health and illness.

Box 3-2 provides an example of scoping for the HIA of a proposed development in Atlanta. The health issues were identified by determining the populations that would be affected and then considering *how* they would be affected. A variety of information was used to inform the process

Establishing Who Might Be Affected

Scoping identifies those likely to be affected by the proposed policy, project, program, or plan. The process may include identifying communities and geographic regions; demographic, economic, racial, and ethnic groups; and vulnerable populations, such as children, elderly people, disabled people, low-income people, racial and ethnic minorities, and people who have pre-existing health conditions. The process of describing pre-existing health issues, health disparities, and influences on health may also begin during scoping, although the full characterization of baseline health status generally takes place during assessment.

54

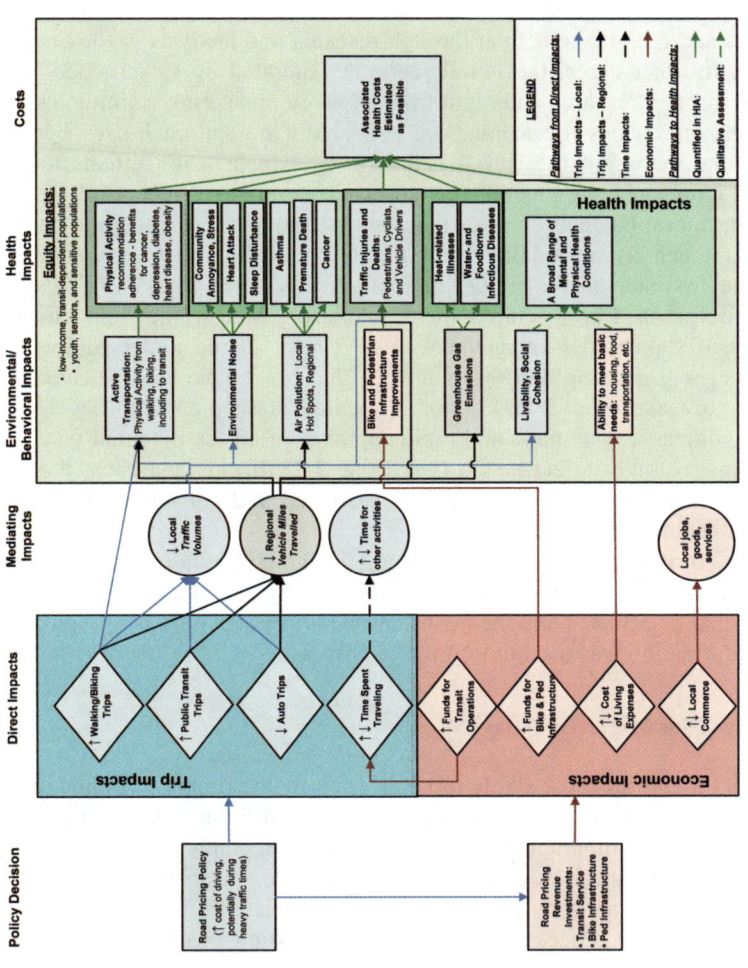

FIGURE 3-1 Example of a logic framework that maps out the possible causal pathways by which health effects might occur. Source: SFDPH 2011.

TABLE 3-1 Example of a Table Used for Systematic Scoping

	Health Category				
Potentially Affected Areas	Chronic Disease	Infectious Disease	Injury	Nutrition	Well-being or Psychosocial
Environment • Air quality • Water quality • Soil • Other					
Economy • Personal (income, employment; can include occupational risk) • Revenue or expense to local, state, or tribal government (support for or drain on services, infrastructure)					
Infrastructure • Need for new roads and transit, water, or sanitation systems • Demand on existing infrastructure					
Services • New services as a direct result of proposal • Drain on existing services resulting from proposed action					
Demographics • Community composition • Traffic volume • Residential or commercial use patterns					
Other					

> **BOX 3-2** Scoping: Atlanta BeltLine HIA
>
> As described by Ross (2007, p. 9), "the Atlanta BeltLine is a transit, trails, parks, and redevelopment project that uses a 22-mile loop of largely abandoned freight rail line that lies between two and four miles from the city center" and affects about 45 neighborhoods. In 2005, an HIA was conducted by a multidisciplinary team. The goal was to incorporate health considerations into the decision-making process "by predicting health consequences, informing decision makers and the public about health impacts, and providing realistic recommendations to prevent or mitigate negative health outcomes" (p. 9).
>
> One of the first steps in identifying the parameters of the assessment was to establish an understanding of the BeltLine, a complex project that had been evolving for several years and was expected to be constructed over a 30-year period. The HIA team needed an authoritative description on which to base its assessment. The Atlanta Development Authority's BeltLine Redevelopment Plan (November 2005) was identified as a coherent and publicly accepted vision that had been approved by local elected officials (ADA 2005). The source of public funding for the project was the Tax Allocation District (TAD), and only within the district's boundaries could funding be collected and bond money spent. A variety of planning and zoning, funding, and environmental regulatory decisions were required for the BeltLine's various components to be developed.
>
> The HIA team was assisted by an advisory committee, but it appears that the committee was not involved in the scoping. The HIA report states that scoping was done by the HIA team and involved desk-based research and a web and postal survey. The scoping phase was used to identify the parameters of the assessment, the affected and most vulnerable populations, and potential key health effects. The final HIA report describes each factor (see below) and presents the results of the scoping. The entire decision-making process is appropriately not described in the final report; however, the reader is not told whether the HIA team prepared a scoping report or whether it presented the findings of the scoping stage.
>
> *Affected populations*: As the TAD constituted only a portion of the city that would be directly affected, the HIA team created the HIA study area by placing a 0.5-mile buffer around the BeltLine TAD. The HIA study area was divided into five planning areas, and census (2000) and mortality data were used to analyze the population profiles. Variations were found in race, age, employment status, poverty, car ownership, and mortality. It was not possible to derive mortality rates for demographic subgroups. Behavioral Risk Factor Surveillance System data were used for the county and the state and stratified by race.
>
> *Most vulnerable populations*: Less information was provided about this step. The most vulnerable populations were identified as people of low economic status, children, older adults, renters, and the carless. Calculations were conducted to develop a vulnerability score. The top 10% of the census tracts within the study area were then identified as locations of the most vulnerable populations.
>
> *Key health effects*: Issues were identified through analysis of newspaper coverage; outreach to such groups as local officials, members of the public, and businesses; development of a logic framework; and a survey of people living, working, or attending school near the BeltLine. The HIA team identified the following critical issues that could affect the health of the study area population: access and social equity, physical activity, safety, social capital, and environment (including air quality, noise, and water management).
>
> Source: Adapted from Ross 2007.

The HIA Team, Advisory Bodies, and Stakeholder Involvement

Scoping also determines who will be part of the HIA team and establishes a plan for technical oversight and review, stakeholder participation and involvement, and involvement of and interaction with decision-makers. Commonly, a core team is responsible for the bulk of writing and analysis. In some cases, the team may draw on outside consultants who have expertise in a specific health issue or method. Furthermore, HIA teams commonly rely on analyses by such experts as traffic-safety engineers or air-quality analysts who provide information on the links between the proposal and changes in health determinants.

Advisory, steering, and technical oversight committees are also commonly convened during scoping. Membership is variable but may include representatives of affected communities or community-based organizations, industrial proponents or business groups, public-health experts, officials involved in the decision-making process, and others who have a stake in the outcome. The committees may be convened for several purposes, including providing technical guidance or peer review, ensuring adequate and fair representation of diverse interests and priorities among stakeholders, communicating the results of the HIA to decision-makers, and developing recommendations that address community needs and are compatible with the specific legal requirements of the decision-making process.

Public and stakeholder participation during scoping can serve several important purposes, such as providing local knowledge regarding existing conditions and potential impacts, introducing alternatives or mitigation measures that stakeholders would endorse as effective ways to address key concerns, and allowing representative participation in shaping the terms of the HIA by groups affected by the proposal. Scoping also establishes a plan for stakeholder participation in later phases of the HIA. The early and central role of stakeholder identification and participation is analogous to the guidance provided in the report published by the Presidential/Congressional Commission on Risk Assessment and Risk Management (1997).

The approaches taken for stakeholder involvement vary widely. The variation partly reflects the wide array of applications of HIA; for example, it is not necessary or feasible to use the same approaches to involve stakeholders for a local project and for a high-level state or national policy. That issue is discussed at greater length in Chapter 4.

Data Sources and Methods

Scoping identifies appropriate data sources for the analysis and should also identify important data gaps. In some cases, the timeline and available resources will prevent collection of new data to address gaps that are identified. In others, scoping may identify studies that can be carried out by the HIA team or studies that can be carried out by experts involved in some other aspect of the planning, permitting, or review process (such as air-quality or traffic-safety analysis). Scoping also establishes a plan for the analytic methods that will be used during the assessment phase. The specific methods used in assessment are discussed in depth in the next section.

Alternatives

Another issue that should be addressed in scoping is identifying alternatives to the proposed action. The cornerstone of an assessment that is conducted to comply with NEPA is the presentation of a set of reasonable alternatives to the proposed action; the assessment then considers the impacts of the proposed action and the alternatives. Assessing alternatives in parallel with the proposal can aid decision-making by highlighting tradeoffs and actions that can be taken to achieve the desired outcome while minimizing harms. Because HIA in the United States is often undertaken outside a formal legal mandate, it has not consistently included alternatives assessment. The committee concludes that when alternatives to the proposal being assessed are under consideration, the HIA team should assess the impacts of each alternative. Because developing an alternative (such as suggesting an alternate route for a proposed highway) involves many considerations that may be outside the purview and expertise of an HIA team, the committee recognizes that it may not be practical to expect the HIA team to develop alternatives independently. However, where practical, the HIA team should aim to evaluate a variety of alternatives or, minimally, to identify the characteristics of proposed actions that would be health-protective or detrimental to health. For example, although an HIA may not be able to incorporate engineering or economic specifications for alternate routes for a proposed highway, it could discuss factors that would influence health outcomes, such as indicating that a desirable route would be, for example, 100 m from any school or elderly facility or would not be proximate to high-population-density areas with a number of vulnerable people. It would then fall to the decision-makers to determine routes that met those criteria.

Outputs of Scoping

On the basis of its review of current guidance and practice, the committee recommends that scoping should result in a framework for the HIA and a written project plan that includes the following:

- An initial brief summary of the pathways through which health could be affected and the health effects to be addressed, including a rationale for how the effects were chosen and an account of any potential health effects that were considered but were not selected and why. Any logic models or scoping tables that were completed should also be included.

- Identification of the population and vulnerable groups—such as children, the elderly, racial or ethnic minorities, low-income people, and communities—that are likely to be affected.
- A description of the research questions, data sources, methods to be used, and any alternatives to be assessed.
- Identification of apparent data gaps and of data collection that could be undertaken to address the gaps or a rationale for not undertaking data collection.
- A summary of how stakeholders were engaged, the main issues that the stakeholders raised, and how they will be addressed or why they will not be addressed.

Committee Conclusions Regarding Scoping

The credibility and relevance of HIA to the decision-making process rest on a balanced and complete examination of the health risks, benefits, and tradeoffs presented by the project, policy, program, or plan being assessed. For that reason, it is important that scoping begin with a systematic consideration of all potential effects rather than limiting consideration to a subset of issues predetermined by the team's research interests or regulatory requirements. Conversely, to have the greatest relevance as an informational and planning document and to ensure judicious use of resources, the HIA should ultimately focus on the health effects of greatest potential importance. Therefore, scoping should be thought of as a two-part process that starts with a systematic effort to identify all potentially important effects and that continues with selection of the most important and well-substantiated effects for further analysis at later stages.

Characteristics unique to the affected community may not be obvious to HIA practitioners who are outside the community. Stakeholders, however, may have insights into local conditions and potential solutions for addressing concerns raised by the proposal under consideration. Scoping should therefore entail a deliberative process that involves engagement of stakeholders. Review of literature and a consideration of the social, economic, and political context of the eventual decision are also important. In selecting the analytic methods that will be used, the HIA practitioner should consider not only technical limitations but what type of information will be most useful to decision-makers.

Finally, it is appropriate to include issues that are the subject of community concern even if they seem unlikely to be substantiated by further analysis. An HIA does not have to accept community concern uncritically. If the HIA is based on a thorough analysis, provides complete information so that community members are able to critique the analysis, and is conducted openly, it may provide reassurance to affected communities even if the conclusions do not support the community's concerns.

Assessment

The assessment phase includes two tasks. The first is to create a profile of the population affected, which includes information on the demographics, baseline health status, and social, economic, and environmental conditions that are important to health. The second task is to analyze and characterize effects on health and its determinants for the proposal and for any alternatives under consideration relative to the baseline and to each other. As part of the assessment phase, a set of specific indicators that can be used to describe the baseline and potential changes in health status or health determinants should be developed. The committee notes that a variety of qualitative and quantitative approaches are often used to generate predictions, but regardless of the methods used, most available guidance emphasizes the importance of considering diverse forms of evidence, a consistent and unbiased approach to selecting and interpreting evidence, and a clear and transparent description of the analytic approach (WHO 1999; Mindell et al. 2008; Fredsgaard et al. 2009; Bhatia et al. 2010).

Given the broad scope of HIA practice settings, applications, and data sources, the committee chose not to develop specific standards or criteria for what constitutes "adequate" evidence or analysis for HIA. Other groups have developed "standard" approaches to promote and evaluate practice quality, including the quality of analysis (see, for example, Fredsgaard et al. 2009; Bhatia et al. 2010). Instead, the committee focused its review on the characterization of effects and the use of evidence, although several recommendations to improve the quality of analysis are discussed in Chapter 4.

Baseline Profile

The baseline profile characterizes the health status of affected populations and includes trends and factors (social, economic, and environmental) known to affect health. Assessing the baseline health status of the affected population provides a reference point with which the predicted changes in health status may be compared; it identifies any groups that could be more vulnerable than the general population to the impacts of the proposal; and it provides an understanding of the factors that are responsible for determining health in the affected communities, and this, in turn, allows for a better understanding of how any changes in those factors may affect health. In general, the baseline profile focuses on health issues and health determinants that may be affected by the proposal rather than on attempting to provide a complete assessment of community health.

Various sources of population-health statistics at the national, state, and local levels are available. But few data may be available on the geographic scale of some decisions addressed in an HIA, such as decisions related to projects that would affect a rural area or a single neighborhood. Moreover, disease rates in small populations can vary substantially, and it may not be possible to calculate them reliably. In such cases, HIAs often rely on data on a larger region and consider whether the characteristics of the larger population can be generalized to the affected community.

Sources of information used in a baseline profile might include census data, hospital-discharge records, disease registries, and population and behavior surveys, such as the Behavioral Risk Factor Surveillance Survey, in which information is collected on different geographic scales. The baseline profile also generally draws on data that describe the social, economic, and environmental conditions important to health, such as labor or housing reports, environmental impact assessments, and gray or unpublished data specific to the decision context.

Developing a robust characterization of baseline health status and the social, economic, and environmental conditions important to health is a challenging aspect of HIA practice. In many cases, a complete and accurate description of health and its determinants in the affected community may not be possible. Some HIAs rely on proxy measures when rates of specific diseases are not available or are too small to calculate. For example, rather than providing an estimate of lung-cancer rates in a small community, an HIA might identify smoking rates and important sources of airborne pollutants in the community's airshed. The committee notes that any limitations, incomplete data, and uncertainty in the baseline analysis should be clearly stated. New surveys to address data gaps or questions specific to the proposal in question are also common in comprehensive HIAs.

Characterization of Effects

Regardless of whether effects are quantified, the assessment stage should include a characterization of each effect to the greatest extent. Different HIA guides vary somewhat as to the specific descriptors that should be used, and practice is highly variable. The core issues that are commonly addressed are as follows:

- *Nature*—describes the effect and the causal pathway.
- *Direction*—indicates whether the effect is adverse or beneficial. In some cases, the direction of the effect may be unclear, or conflicting influences on a given health outcome may be identified (Harris et al. 2007).
- *Intensity*—indicates the severity of the effect (for example, fatal, disabling, or no disability).
- *Magnitude*—refers to the expected size of the effect and can be described by the number of people affected or by expected changes in the frequency or prevalence of symptoms, illness, or injury.
- *Distribution*—delineates the spatial and temporal boundaries of the effect and identifies various groups or communities that are likely to bear differential effects. This factor is important for ensuring that health equity is addressed.

Groups can be defined by age, sex, ethnicity, socioeconomic position, locational disadvantage, and health status or disability (Harris et al. 2007). Recognizing and addressing the effects of a proposal on health equity (or health disparities) between various groups has been seen as a core task of HIA, although HIA practice has sometimes been criticized for a lack of attention to health equity (Scott-Samuel 1996; WHO 1999; Harris et al. 2007).

- *Timing and duration*—indicates at what point of the proposed activity (such as construction vs operation of a new power plant) the effect will occur, how long it will last, and how rapidly the changes will occur; also discusses whether effects are reversible or permanent.
- *Likelihood*—refers to the chance or probability that the effect will occur.
- *Confidence or certainty*—characterizes the effect according to level of confidence or certainty in the prediction; that characterization is based on the strength of the evidence as described below.

Some HIA guides recommend using a matrix, such as those shown in Tables 3-2 and 3-3, to characterize effects (Harris et al. 2007; ICMM 2010). A matrix can be a useful way to organize a qualitative analysis and to convey results in a manner that is easy to understand, but a matrix may also be misinterpreted as being more objective than a simple description. It is important to note that a matrix does not explain how evidence was used to reach conclusions. A clear explanation should be provided with the characterization of effects that indicates the evidence used to develop the matrix and any limitations, data gaps, and uncertainties.

The committee notes that in addition to characterizing effects, HIAs may provide conclusions on the significance—or societal importance—of the effects, although this has been rare in U.S. practice. Assignment of significance rests on the characterization of an effect as described above, but judgments regarding what constitutes a significant impact are ultimately determined partly on the basis of social and political values.

Evidence and Approaches

Characterization of health effects in HIA relies on qualitative and quantitative evidence. The wide array of evidence includes public testimony on local conditions and concerns, interviews with key informants, surveys, epidemiologic analysis (for example, observational cross-sectional surveys, longitudinal studies, and intervention or experimental studies), measurement of physical environmental conditions and modeling (for example, modeling of infectious-disease propagation or dispersion of noise and air pollutants), and expert opinion. In many cases, the first course of action is to gather information from published literature, unpublished reports, administrative data gathered for routine

TABLE 3-2 Example of a Matrix for Analyzing Health Effects

Main Health Outcome or Health Determinant	Health Outcome or Health Determinant Sub-Category	What is the health impact? Positive, negative, uncertain or no effect	Who benefits? Who is negatively affected? Who is unaffected? Whole population, vulnerable group Heath equity Cumulative impacts	Pathway of health impact How does the impact occur	Magnitude/consequence of impact Low Medium High	Likelihood/probability of impact Low (possible) Medium (probable) High (definite)	Significance of impact (Magnitude × Likelihood) Low Medium High	Degree of Confidence of the impact occurring Low Medium High
Health Outcome								
	Infectious disease							
	Chronic disease							
	Nutritional disorders							
	Injury							
	Mental health and well-being							
Health Determinant								
Individual/family	Physiological							
	Behaviour							
	Socio-economic circumstances							
Environmental	Physical							
	Social							
	Economic							
Institutional	Organization of health care							
	Other institutions							
	Policies							

Source: ICMM 2010. Reprinted with permission; copyright 2010, International Council on Mining and Metals.

TABLE 3-3 Example of a Table for Rating Importance of Health Effects

Magnitude of impact		Likelihood of Occurrence of a Health Impact		
		Low	Medium	High
Health impact rating	Description	Unlikely to occur	Likely to occur sometimes	Likely to occur often
0	None	No significance	No significance	No significance
1	Low	Very low significance	Low significance	Medium significance
2	Medium	Low significance	Medium significance	High significance
3	High	Medium significance	High significance	High significance

Source: ICMM 2010. Reprinted with permission; copyright 2010, International Council on Mining and Metals.

monitoring purposes, and other available documents. Such reviews of the existing literature are common. The degree to which they are systematic varies, and some authors have suggested frameworks and guidance for conducting systematic reviews (Mindell et al. 2004, 2010). The available data, however, may not be sufficient, and the HIA team may make a decision to dedicate resources to collection of new data. The use of various types of evidence and approaches used to make predictions are discussed below. The committee notes that each approach for gathering and interpreting data may be conducted in ways that are more or less participatory, engaging stakeholders in shaping the research questions, interpreting the findings, and developing recommendations on the basis of the findings.

Qualitative evidence provides a context-specific view of people's lives. Qualitative data can be gathered through, for example, focus groups, one-on-one interviews, surveys, individual meetings with stakeholder organizations, testimony in community meetings, Web-based or other written input, and running a stand or exhibition in a public place. Participatory approaches that actively engage stakeholders in the process can yield rich information and provide opportunities for stakeholders—including community members—to influence the questions asked and to participate in the interpretation of findings. The approaches can provide useful information on how people view the proposal, that is, how it is expected to affect them and potentially improve or harm their quality of life. A central tenet is that people's experience offers an invaluable perspective on the potential effects of the proposal.

The selection of qualitative and descriptive approaches will be informed by the scale and size of the proposal, the profile of the affected population, and the uses of the resulting information. Qualitative approaches can more easily present the causal pathways in terms used by participants; this ensures that different voices are presented in the HIA and can increase the legitimacy and stakeholder's ownership of the process and results. Box 3-3 provides an example of an assessment step that was based on interviews with people who were likely to be affected by a decision and that considered their impressions of the effects that industrial activities were having and were likely to continue to have on individual, family, and community life. The committee notes that qualitative social sciences and participatory-action research provide fertile ground for innovative methods for HIA. However, it is important to note that the use of qualitative approaches in HIA should not be interpreted as implying a need for less analytic rigor. As for any other research method, qualitative analysis in HIA should use appropriate methods and a clear, rigorous research design.

BOX 3-3 Assessment: Northeast National Petroleum Reserve-Alaska

In 1998, the Bureau of Land Management (BLM) completed a Northeast National Petroleum Reserve-Alaska Integrated Activity Plan/Environmental Impact Statement (EIS). BLM later considered amending the EIS to make additional public lands available for oil and gas leasing in the original 4.6-million-acre planning area. The local North Slope Borough government—an agency that participated in preparing the EIS—conducted an HIA, which was integrated into the EIS report.

The HIA drew on public testimony, literature review, and accepted mechanisms of health and illness to establish the scope of health concerns that should be considered. A logic framework was produced to guide the assessment. The associations between expected disturbances and changes in health were then analyzed in more depth to look at the alternatives proposed for the expansion and at the cumulative effects of oil exploration and extraction. The HIA team identified the pathways in which the expansion would affect the lives of the primarily Iñupiat residents of the area. Activities directly or indirectly associated with oil and gas—including aircraft traffic, seismic exploration, influx of nonresident workers, and emissions and discharges—were identified, and pathways were analyzed to consider their impacts on such problems as diet-related chronic illnesses (for example, diabetes and hypertension), food insecurity, and social pathology (for example, assault, alcohol and drug abuse, and violence). Those pathways and others were evaluated by using public-health data, literature on analogous populations, knowledge about accepted mechanisms of health and disease, witness testimony, and the effects analysis of other resources.

The discussion described pathways by which Iñupiat health was likely to be affected. For example, under Alternative A, diabetes and metabolic disorders would be expected to increase if impacts on subsistence led to declining subsistence harvests (through declining populations of subsistence resources, displacement of resources that made hunting less successful, or displacement of hunters by oil activity and infrastructure). It identified which areas and villages were most likely to be affected and when. The assessment also highlighted several potential benefits, such as "funding for infrastructure and health care; increased employment and income; and continued funding of existing infrastructure" (BLM 2007, p. 500). Because the biologists on the EIS team were uncertain of the degree to which subsistence harvests might be favorably or adversely affected, the HIA did not attempt to make quantitative estimates of the probability or intensity of the impact.

Public-health professionals reviewed the results of the analysis. On the basis of the findings, the HIA recommended a series of public-health mitigation measures that were selected to maximize any favorable impacts and to minimize harms.

Source: Adapted from BLM 2007 and Wernham 2007.

Quantitative evidence can include routinely collected information, such as mortality and census data, that can inform the baseline assessment. It can also include information from social-science and epidemiologic studies regarding the strength of associations between the social and physical environment (such as air and water quality and economic impacts) and health outcomes—information essential in the quantitative prediction of health effects. A large and growing body of quantitative evidence is available; where specific additional information is needed and resources are available, HIA teams may collect new quantitative data in the course of conducting an assessment.

If there is a causal relationship between variables, a valid estimate of effect size, and data on how a decision will change the prevalence of a health-related factor (exposure), it is possible to make quantitative predictions of effects (Fehr 1999; Veerman et al. 2005; Bhatia and Seto 2011). Potential health effects have been estimated by using established approaches for quantitative analysis, such as the calculation of the fraction of disease rates in a population that can be attributed to the risk being analyzed and the application of available exposure-response functions to quantify cancer risk associated with incremental changes in exposure to carcinogens. Additional modeling approaches, such as system-dynamic modeling and agent-based models, are also promising and emerging tools that could have applications to health. Box 3-4 provides several examples of topics that have been addressed in HIA by using quantitative methods.

Outputs of Assessment

Assessment should result in a report that

- Describes the baseline health status of the affected population with appropriate indicators, including prevalent health problems, health disparities, and social, economic, and environmental factors that affect health. The baseline should be focused on the issues that are likely to be affected by the proposal.
- Analyzes beneficial and adverse health effects and characterizes the changes in the indicators selected, to the extent possible, in terms of nature, direction, intensity, magnitude, distribution in the population, timing and duration, and likelihood.
- Integrates stakeholder input into the analysis of effects.
- Describes data sources and analytic methods and methods used to engage stakeholders.
- Identifies limitations and uncertainties clearly.

Committee Conclusions Regarding Assessment

The selection of analytic methods for HIA is driven by the complex pathways and the multiple, sometimes conflicting, influences on any given health outcome and is also affected by the decision-making context. Decision-making is rarely based solely on scientific evidence but instead takes into account an array of political, economic, technical, and practical considerations. Decisions are often based on incomplete information and must often be made within a specified time rather than waiting for more complete information. By necessity, therefore, impact assessment is a pragmatic exercise and reflects a balance between scientific rigor and professional judgment. Expert judgment is central to HIA but must be grounded in a solid foundation of scientific neutrality and accepted public-health principles. An explicit statement of data sources, methods, assumptions, and uncertainty is essential, but uncertainty does not negate the value of the information. Even when there is substantial uncertainty, an assessment can illuminate potential causal pathways that—even when there appear to be conflicting influences on a specific outcome—can point the way toward a flexible framework for monitoring and managing any impacts that might occur as the proposal is implemented.

Literature review provides much of the empiric evidence for most HIAs, and whenever possible, assessors should conduct a systematic review of the literature for any health effects and determinants identified as high-priority issues in scoping. Failing to undertake a systematic review may mean overlooking evidence that would lead an assessor to a different conclusion. In practice, not all HIAs have conducted systematic literature reviews or documented review methods. If it is not possible to undertake complete, systematic literature reviews for an effect analyzed in an HIA, HIA practitioners must be vigilant to avoid selective searching and consideration of only studies that confirm particular conclusions (Mindell et al. 2004). However the literature review is conducted, the methods should be clearly described in the report, and any studies with conflicting results acknowledged.

BOX 3-4 Examples of Health and Behavioral Effects That Have Been Addressed Quantitatively in HIA

The bulleted list below provides examples in which some health impacts or behavioral outcomes have been quantified. The committee notes that in any assessment, it would be difficult or impossible to quantify all potential health impacts.

- Cancer risk associated with exposure to hazardous chemicals.
- Shortened life expectancy associated with air pollution.
- Injuries and fatalities associated with changes in vehicle traffic or speeds.
- Physical activity associated with changes in pedestrian infrastructure.
- Alcohol-consumption effects of alcohol taxes.
- Cancer risk and changes in life expectancy associated with tobacco taxes.
- HIV-AIDS infection risk associated with oil-pipeline construction.
- Life expectancy and physical function associated with income.

Sources: Veerman et al. 2005 and Bhatia and Seto 2011.

The reliability and validity of predictions made in HIAs have often been questioned (Thomson 2008). That issue will probably continue to challenge the credibility of HIA practice in the eyes of some audiences and highlights the need for continued research and refinement of methods to improve its value to decision-makers (Petticrew et al. 2006). Issues surrounding uncertainty, literature review, and reliability and validity of predictions are discussed in greater depth in Chapter 4.

Recommendations

Recommendations identify specific actions that could be taken to avoid, minimize, or mitigate harmful effects identified during the course of the HIA or to take maximal advantage of opportunities for a proposal to improve health. Depending on the nature of the proposal being assessed and the specific impacts, recommendations can take various forms (see Box 3-5), such as the following:

- A major alternative to a proposal (for example, routing a proposed highway away from a vulnerable population or building a light-rail line rather than widening a road).
- Mitigation measures that address a specific impact identified in the HIA and are intended to minimize a potential harm (for example, a measure to reduce benzene emissions from gas wells near residential areas) or measures to maximize a potential benefit.
- Health-supportive measures that would generally support health but are not tied directly to a specific impact (for example, building a clinic in an underserved neighborhood that would be adversely affected by emissions from a new freeway).
- Adopting a position for or against a proposal (for example, support for or opposition to a legislative proposal).

An HIA, however, might not provide any recommendations; this could occur if the HIA failed to reveal any important health effects. In some cases, the lack of a recommendation may reflect a desire to avoid a perception that the HIA is a one-sided advocacy exercise, particularly when options for recommendations would involve adopting a position wholly supportive of or opposed to the proposal being assessed.

The development of recommendations should be guided by a consideration of any available evidence regarding effectiveness. Such evidence may come from a review of published literature on interventions to address the health outcomes of concern. Or, in some cases, there may be unpublished evaluations of measures that have been implemented in similar scenarios. However, because few studies have directly assessed the impact of the implementation of policies, plans, programs, or projects on health outcomes, there may be little direct evidence available with which to predict a given measure's effectiveness. In those cases, the HIA team may need to rely on established principles of health promotion and disease prevention to develop approaches to minimizing or mitigating the identified effects. The committee emphasizes that the effectiveness of recommendations depends not only on the scientific validity of the interventions identified but on their relevance to the affected community's concerns and their applicability within the regulatory or legislative framework of the proposal being considered. Chapter 4 discusses the extent to which an HIA can ensure the implementation of recommendations.

BOX 3-5 HIA Recommendations

HIA recommendations take various forms, and some examples are provided below. The committee is not endorsing the HIAs or the recommendations, but simply providing examples.

- *Alternative to a proposal.* As described in Box 3-3, the HIA of oil and gas leasing in the National Petroleum Reserve-Alaska raised concerns regarding the potential for adverse effects on the culture, well-being, and health of local residents because of the risk of disrupting the local fish and game on which the community depends for food. All three leasing alternatives presented in the environmental impact statement raised similar concerns. To address the concerns, the North Slope Borough suggested restrictions on leasing in a small percentage of the area. The final decision by the Bureau of Land Management reflected a consideration of those concerns and deferred leasing in the most critical fishing and hunting areas, which represented a small percentage of the total area available for leasing (BLM 2008).
- *Mitigation measures.* An HIA of rezoning from industrial to residential use in San Francisco—a plan that would add 30,000 households—identified health-related noise and air-quality issues for the proposed residential units. It recommended new standards for ventilation and acoustical protection for new development. As a result, the city adopted performance-based regulations to ensure indoor-air quality and noise protections for all new residential development (Bhatia and Wernham 2008).
- *Mitigation and health-supportive measures.* An HIA of proposed oil development in Sakhalin Island, Russia, concluded that a large influx of oil and gas workers from outside the region could increase the risk of sexually transmitted illnesses in workers and the community. The HIA proposed mitigation measures (such as restricting access to the work camp by local residents) and health-supportive measures (such as "supporting the health community in improving STD programme management") (Balint et al. 2003).
- *Adopting a position for or against a proposal.* An HIA of proposed restrictions in the funding for the Massachusetts rental-voucher program for low-income residents found that the restrictions could be harmful to health and recommended against them (Child Health Impact Working Group 2005).

Recommendations are often developed throughout the HIA process. It is common for mitigation measures and design alternatives to be considered during scoping, to be refined as the assessment phase further characterizes the impacts and identifies their importance, and to be made final during the recommendations phase. The process is analogous to the approach in the new risk-based decision-making framework proposed in *Science and Decisions* (NRC 2009), in which the primary objective of risk assessment is to help decision-makers choose among risk-management options by providing information on health risks that can be considered in the context of economic, social, and other factors. Similarly, HIA recommendations concern measures that can be taken to protect or improve health, but ultimately the decision-makers must weigh those recommendations with the political, economic, social, and technical factors that are relevant to the decision. In some cases, recommendations are developed by a decision-maker in response to an HIA report (Quigley et al. 2006). As discussed later in this section, recommendations can also establish a foundation for monitoring, and the results of the monitoring may indicate that the management strategies need to be adapted to respond to the observed outcomes—a process known as adaptive management (Johnson 1999).

The Roles of the Public and Decision-Makers

Public input while recommendations are being developed helps to ensure that proposed measures are locally relevant, address context-specific factors that might render them more or less effective, and address public concerns and hopes. The success of recommendations ultimately depends on the public's trust in and support of them. For example, in Alaska, one of the adverse impacts of a proposed mine expansion was the feared contamination of water and wildlife, and evidence suggested that a fear of contamination might lead communities to shy away from eating a traditional diet. To address that concern, monitoring of concentrations of selected contaminants in local fish was proposed as a mitigation measure. Community input on the proposal suggested that for the program to reassure community members effectively, the monitoring should be conducted by an independent third party, and there should be strong community oversight at each stage.

Because decision-makers must eventually translate health-based recommendations into actionable measures (for example, by modifying legislation, drafting regulations or permit conditions, instituting new zoning requirements, or encouraging voluntary activities), regular communication between the HIA team and the decision-makers is important for the success of a proposed recommendation (EPA 2009). As in the realm of health risk assessment, there remains a need to distinguish between the assessment and management phases to avoid manipulation of analytic components by decision-makers. However, *Science and Decisions* (NRC 2009) emphasizes that a detailed understanding of the decision context is necessary for analyses to be scoped appropriately and that the conceptual distinction between assessment and management should not be interpreted as a firewall that prevents communication between parties. Having transparency throughout the process and clearly delineating the roles and responsibilities among various parties will help to limit real and perceived bias. Mechanisms to limit bias in decision-relevant analyses further are discussed in Chapter 4.

Health-Management Plan

HIA guidance often points out the need for monitoring and continuing management and verification that mitigation measures are being implemented. A plan for continuous monitoring, adaptation of mitigation measures, and verification of performance—although not currently a uniform aspect of HIA practice—helps to ensure that measures are carried out and achieving their objectives. Such a plan is often referred to as a public-health management plan or a health-action plan (Quigley et al. 2006). Recommendations form the core of a health-management plan, but the plan also determines authority for and assigns responsibility for implementing each recommendation, establishes a monitoring plan, and creates or suggests mechanisms to verify that assigned responsibilities are being met. Monitoring focuses on measures that are likely to be sensitive and early indicators of change. Selection of appropriate indicators will be discussed at greater length below in the section "Monitoring and Evaluation."

The health-management plan suggests which stakeholder agency or entity could take responsibility for implementing each recommendation. Recommendations may be implemented through regulatory mandates or voluntary actions by stakeholders. Industrial proponents, government decision-making agencies, local health departments, and independent organizations (such as universities and nongovernment organizations) may all be in a position to implement measures recommended in the HIA.

Management of the health effects of a proposal as it moves from planning into implementation should be a dynamic process in which monitoring results may drive continued adaptation of the health-management plan. As noted above, the iterative process is known as adaptive management in the field of environmental management.

Outputs of Recommendations

The recommendations should be provided in the final HIA report and should document available supporting evidence, stakeholder input, and a health-management plan, which should do the following:

- Discuss what entity has the authority or ability to implement each measure and document any commitments to do so.
- Propose appropriate indicators for monitoring.

- Propose a system to verify that measures are being implemented as planned.

If no recommendations are made in the HIA report, an explicit rationale should be provided for the decision not to include them.

Committee Conclusions Regarding Recommendations

Making recommendations is a well-accepted part of HIA practice, but relatively little attention has been paid to how they should be formulated. The committee notes three considerations that may be particularly important for producing effective, actionable recommendations. First, community input is essential especially for proposals that will affect the local community primarily. Community input during the development of recommendations can ensure that they address specific aspects of living conditions and community design that may not be obvious to an outside researcher, and it provides an opportunity to ensure that the recommendations address high-priority issues in a manner that is acceptable to the affected community.

Second, recommendations are effective only if they are adopted and implemented. Adoption of recommendations depends partly on the involvement of decision-makers in the HIA process (Elliot and Francis 2005; Davenport et al. 2006). A gulf may exist between an intervention that is sound from a public-health perspective and one that is acceptable and can be acted on within the relevant regulatory or legal framework. Drafting measures that address identified public-health risks and that fulfill the requirements of the legal framework governing a decision will increase the chances that HIA recommendations are implemented. Drafting measures that can be readily incorporated into statutes, regulations, zoning provisions, or permit conditions with little adaptation may also increase chances of implementation. Collaboration with decision-makers or consultation with experts familiar with the legal or regulatory context may be the most effective way to ensure that recommendations are pragmatic and can be practically incorporated into the decision-making process.

Third, recommendations should include the elements of a health-management plan, including a consideration of appropriate indicators for monitoring, identification of entities that have the authority or ability to implement each measure, and a mechanism for verifying implementation and compliance. That permits recommendations to form the basis of effective implementation and management rather than merely providing a static system without the capacity to adapt. The process of implementing recommendations should be transparent and should include opportunities for public participation in the decision process and clear mechanisms of accountability.

As a final note, it is important to remember the context in which HIAs are conducted when considering the recommendations phase. As discussed at the beginning of this chapter, HIAs in the United States are often conducted without a formal legal mandate and by an agency or organization that does not have decision-making authority. In practice, therefore, the HIA team will be asking a decision-maker to consider the findings and recommendations. The decision-maker must ultimately balance health considerations with the many technical, social, political, and economic concerns that bear on the proposal. The use of the information by the decision-maker is discussed at greater length in Chapter 4 in the section "Managing Expectations."

Reporting

Reporting is the communication of the findings and recommendations of an HIA to decision-makers, the public, and other stakeholders. It includes the production and dissemination of written materials that document the HIA process, methods, findings, recommendations, and limitations of the analysis; and it includes the public dissemination of results through other channels, such as meetings with the public, decision-makers, and other stakeholders. The information generated by the HIA process needs to be organized and presented in such a way that it can be readily understood by the intended audiences and present a compelling case for recommended actions. Box 3-6 shows how the results of an HIA of proposals to provide paid sick days to employees were presented clearly in a report with appropriate acknowledgement of the strengths and weaknesses of the evidence. It also shows how HIA results can be disseminated widely in different formats through a number of channels. For example, HIA reports can be disseminated in hard copy, in electronic format, at public meetings, to focus groups, or at different stages in the HIA process or policy cycle. The committee notes that effective dissemination requires consideration of barriers—including those associated with language, availability of child care, disability, access to transportation, disenfranchisement, or literacy—and that multiple approaches may be required for disseminating a single HIA so that all appropriate audiences can be reached. That issue is addressed again in Chapter 4.

In some cases, the HIA process allows a period for formal public comment on a draft of the HIA report. The final draft responds to public comments and incorporates necessary changes or new information. The process mirrors the one set out by NEPA for an environmental impact statement, but the practice is far more variable for HIA. In other cases, a draft may be submitted to an internal body, such as a steering group, whose comments are incorporated into a final public version.

In practice, however, reporting may occur at earlier stages of the HIA process and include public meetings; meetings with decision-makers, other stakeholders, and advisers; and dissemination of interim public reports, such as a scoping summary. HIA is meant to assist decision-makers, so although the act of reporting is a formal step in the HIA process, it is also in the interest of decision-makers and the HIA team to keep in constant communication throughout the HIA process so that emerging results can be incorporated into the policy, plan, program, or project.

> **BOX 3-6** Reporting: Legislation on Paid Sick Days
>
> The National Partnership for Women and Families commissioned Human Impact Partners and researchers at the San Francisco Department of Public Health to conduct an HIA of the federal Health Families Act of 2009, which would guarantee workers access to paid sick leave. The research was funded by the Annie E. Casey Foundation as an initiative with the potential to encourage long-term strategies and partnerships to strengthen families and communities. Human Impact Partners then worked with groups in other states to extrapolate the findings of the national report to local jurisdictions to analyze the health effects of paid sick days.
>
> The report of the Healthy Families Act HIA provides a clear description of the steps in the analytic process. The key findings are provided in the opening section of the report, and they are categorized according to the strength of the evidence as "highly likely," "likely but less well-supported by the available evidence," and "plausible, but not well-supported." For example, according to the report, a requirement for paid sick days is highly likely to lead to more workers taking leave to recuperate from an illness, to receive preventive care, or to care for ill children and dependents. It is also highly likely to lead to "improved compliance with public-health guidance regarding seasonal influenza and community mitigation strategies for pandemic influenza." In contrast, effects that are likely but less well supported include increased ambulatory or preventive primary care, fewer emergency-room visits by workers who are insured, and greater compliance with infection-control policies. Finally, effects that are plausible but are not supported by available evidence include fewer hospitalizations because workers are able to receive the preventive primary care needed to maintain good health.
>
> The results of the HIA were presented in different formats; the full report was accompanied by a summary and fact sheets. The findings of the HIA were covered by newspapers and Web sites in California, Maine, Massachusetts, and New Hampshire; and the HIA researchers were interviewed on radio. The press coverage recognized the tension between the burden that this new requirement would place on businesses and how the health of employees and the wider community are affected by people who work while they are ill. Human Impact Partners noted that many—including labor groups and funders—used the HIAs to assess work and family issues. The HIAs also changed the debate in such a way that providing paid sick days for employees began to be presented as a public-health issue rather than a labor issue. For example, the chair of the California Assembly Labor Committee referred to the HIA and "asked the opposition to the bill if they condoned the spread of disease through restaurant workers."
>
> Sources: Adapted from Cook et al. 2009; Human Impact Partners 2009a,b; AECF 2011.

The quality of the report can be a criterion by which the quality of the process is judged; that is, How clearly does the final document present the results of the analysis? It is critical to arrange the information logically so that readers can navigate easily through the document, to provide a lay summary that accurately describes the main findings and conclusions of the study, and to reference all data and sources accurately (Fredsgaard et al. 2009).

Transparency of HIA

HIAs in the private sector are increasingly common, pursuant to internal corporate guidelines or requirements of lending banks, such as the International Finance Corporation and World Bank (see Appendix A for further discussion) (Birley 2005; IPIECA/OGP 2005; McHugh et al 2006; ICMM 2010; IFC 2007, 2010).[3] Few, however, are made public. Disclosure requirements and practices vary considerably among development lenders and private-sector proponents. The World Bank and International Finance Corporation have policies governing the disclosure of information, and although the policies differ, both provide for withholding or excluding documents that might contain proprietary information or information whose disclosure could damage a client or lender's financial, political, or legal interests (Halifax Initiative Coalition 2006; IFC 2006, 2010; McHugh et al. 2006; World Bank 2010). For private corporations undertaking an HIA, the decision of whether to make an HIA public and what to disclose may be governed by internal corporate policies, by the standards of lenders supporting the project, or by a government that has jurisdiction over the project (McHugh et al. 2006). A number of corporations and professional associations, such as the International Committee on Mining and Metals and the International Association of Oil and Gas Producers, have guidance for HIA, but relatively few completed industry-led HIAs or environmental, social, and health impact assessments are available on the Internet or on public Web sites that catalog HIA activity.

A related issue is incomplete disclosure—such as disclosure of only summary information without data or analysis, disclosure only by electronic media in communities unlikely to have access, and English-only reports. Incomplete disclosure may substantially limit access to complete information regarding the process, data sources, methods, and findings of an HIA for those who will be affected by the proposal being assessed (McHugh et al. 2006).

Failure to disclose HIA results and incomplete disclosure are not restricted to industry. Public agencies might not disclose or might redact or otherwise limit disclosure of information. Similarly, HIAs sponsored by private nonprofit organizations may not have requirements for disclosure inasmuch as most U.S. HIAs are not done under a legal mandate that requires disclosure. However, many HIA reports are available from public agencies, universities, and nonprofit organizations, and the committee found few examples of HIAs led or commis-

---

[3]The committee is referring here to HIAs sponsored or led by private-sector entities that are *not* part of any formal government process, such as a permitting or regulatory requirement. HIAs conducted as part of a formal government process are generally subject to disclosure and freedom-of-information requirements.

sioned by the private sector that were available. Given that HIA led by the private sector appears to be a rapidly increasing practice, the issue of availability bears further consideration.

Outputs of Reporting

The final HIA report should document the following:

- The nature of the proposal being assessed, including alternatives that were included in the analysis.
- The population, subgroups, vulnerable populations, and stakeholders likely to be affected and how they were involved in the HIA process.
- Data sources and analytic tools used.
- Findings of each stage of the HIA and a summary of outputs at the end of each stage.

In addition to a final report, stand-alone executive summaries or fact sheets can help to disseminate and communicate the findings and recommendations of an HIA to various key audiences.

Committee Conclusions Regarding Reporting

Across the field, there is little uniformity in the content of written HIA reports. The committee finds that an HIA report should at least describe the proposal and alternatives that are the subject of the HIA, the data sources and analytic methods used, the groups and individuals that were consulted in the course of the HIA, the process and findings of each step of the HIA, and the overall conclusions and recommendations. The HIA conclusions and recommendations should be presented in a manner that is clear and easily understood.

The committee recommends that HIAs be publicly released and disseminated. Although little has been written on the reasons for keeping HIA information confidential, the committee recognizes that there may be reasons for organizations conducting HIAs to decide not to disclose the results. For example, there may be concerns about risks to a proponent's reputation or to the viability and public acceptance of a proposed project if a report discloses important unmitigated adverse impacts or potential impacts that are uncertain or for which strong evidence does not exist. There could also be concerns that disclosure of such information would lead to litigation. Furthermore, impact assessments, including HIAs, may rely on proprietary business information whose disclosure is legally barred or could damage a proponent's business edge or competitiveness.

Notwithstanding those considerations, the committee considers the public disclosure of HIAs to be an important ideal of practice but recognizes that it may not be realistic to expect widespread disclosure in the absence of requirements or incentives for it. However, the committee notes that there are several benefits of disclosure for industry, policy-makers, and the affected communities. First, disclosure informs affected communities and individuals and possibly other stakeholders, such as government agencies and officials, of possible effects on their health and well-being, a core objective of HIA. Second, it allows findings to be reviewed and improved. Third, it informs government agencies and officials of potential changes in demand for services, such as health care, emergency response, and public safety; this can facilitate an appropriate response. Fourth, disclosure of potential impacts may benefit industry by reducing the risk of litigation and by reducing tort liability by fulfilling requirements to warn those potentially responsible and potentially affected before the effects occur. Fifth, transparent reporting of possible environmental and health impacts has proved in many studies to lead to risk reduction because it motivates changes, such as improved pollution controls, on the part of industry and governments (Wolf 1996; Bennear and Olmstead 2008; Vaccaro and Madsen 2009). Sixth, because many established environmental risk factors are found at higher concentrations in vulnerable communities, disclosure of risks may be an important way to reduce health disparities and address concerns about environmental justice (Miranda et al. 2008). Seventh, disclosure allows people to take voluntary actions to avoid risk (Neidell 2009). For those reasons, the committee concludes that HIAs—including, to the extent practical, the data used for the analysis, analytic methods, assumptions, findings, uncertainties, data gaps, and recommendations—should be made public.

A well-designed dissemination strategy is critical for the success of an HIA. The dissemination strategy should be developed in a systematic manner, should consider what groups need or will rely on the information (including stakeholders and decision-makers), and should determine the most effective ways to present the information to these groups, taking into account any barriers or challenges.

Simply producing and disseminating a report may not be sufficient to secure adoption and implementation of HIA recommendations. Robust and continuing efforts to inform decision-makers of the findings and recommendations of the HIA and efforts by HIA practitioners and other stakeholders to champion choices that will benefit health can be an essential part of an effective HIA. Available studies suggest that efforts to involve and inform decision-makers throughout the HIA process and a strong relationship between the HIA team and decision-makers are often critical for the HIA's effectiveness (Veerman et al. 2005; Morgan 2011). It is critical for the credibility of the HIA that the measures or outcomes being promoted are grounded in full and transparent consideration of the evidence that supports and does not support the issue in question. Efforts to support health-based recommendations must be carefully distinguished from biased efforts to promote a specific outcome or measure on the basis of an incomplete or inaccurately weighted comparison of favorable and unfavorable aspects of a proposal or of a predetermined political agenda. The committee recognizes that undue bias in an HIA may compromise its credibility and efficacy.

Monitoring and Evaluation

Monitoring and evaluation are often, although variably, described as the final stage of HIA (see Appendix E). Some have suggested that evaluation should be considered as outside the HIA process itself because of the need for an independent and objective perspective, particularly for impact evaluation (Bhatia et al. 2009). Several types of evaluation may be conducted on an HIA, including the following:

- *Process evaluation.* Considers whether the HIA was carried out according to the plan of action and applicable standards.
- *Impact evaluation.* Seeks to understand the impact of the HIA itself on the decision-making process or on other factors outside the specific decision being considered.
- *Outcome evaluation.* Focuses on the changes in health status or health indicators resulting from implementation of the proposal.

In practice, most HIAs do not include process, impact, or outcome evaluation; this has been attributed to a lack of interest, time, and resources in the case of process and impact evaluation and to the length of time (often many years) required for observing changes related to implementation. The discussion below briefly provides definitions and key features of HIA monitoring and evaluation.

Monitoring

As previously described in the section on "Recommendations," monitoring can refer to tracking changes in health indicators as a new project or policy is implemented and has been defined as outcome monitoring. Indicators may be health outcomes in some cases, whereas health determinants may be more appropriate in others. For example, if a traffic-calming infrastructure was installed on a street that had a high rate of pedestrian injury, it may be appropriate to monitor injury rates directly because changes would be expected as soon as the installation was complete. In contrast, the effect of decisions on some health outcomes (such as cancer or obesity) may take years to occur and may have multiple contributing factors. In those cases, it may be more appropriate to monitor exposures—such as environmental concentrations of a carcinogen or the availability of safe walking corridors—that are linked to the outcome of interest by public-health evidence.

Process Evaluation

Process evaluation assesses the design and execution of the HIA in light of its intended purpose and plan of action and applicable practice standards. Process evaluation can range from a simple self-assessment that is undertaken at the end of an HIA and focuses on a few variables that are relatively simple to describe, track, or measure—such as the methods used, degree of certainty of predictions, and approach to stakeholder engagement—to a more comprehensive case study that seeks to evaluate the HIA process holistically. Observing and documenting the HIA process—such as methods of engaging stakeholders and interacting with decision-makers and approaches to addressing analytic challenges—and interviewing participants and stakeholders are the main methods of process evaluation.

Impact Evaluation

Impact evaluation attempts to judge whether the HIA influenced the decision-making process, that is, whether and to what degree the recommendations were adopted and implemented and how the HIA influenced the decision-making process. It can also assess whether the HIA had other important effects, such as building new collaborations among agencies, ensuring that stakeholder perspectives were considered, and increasing awareness of previously unrecognized health considerations. In some cases, the impact of the HIA on a decision is clear-cut. For example, in the Alaskan oil and gas HIA mentioned in Box 3-3, the HIA team drafted recommendations in collaboration with the decision-maker, the Bureau of Land Management, which formally adopted the recommendations as mitigation measures.

In other cases, it may not be possible to attribute a particular decision to the influence of an HIA (Wismar et al. 2007). For example, in Oregon, an independent health-oriented nonprofit organization conducted an HIA of a series of proposals to reduce vehicle miles traveled in a bill intended to reduce greenhouse-gas emissions (UPH 2009). The enacted legislation is consistent with some of the recommendations of the HIA, but there were no data to evaluate whether those drafting the legislation were influenced by the recommendations; there were no interviews with legislators over the course of the legislative process (Human Impact Partners 2010). Observations that might indicate some influence of the HIA include discussion about HIA by legislators debating a proposal. In that case, a robust evaluation method, such as interviews conducted with decision-makers before and after the HIA, could provide the data needed to gauge the effect on decisions.

Impact evaluation can also help to determine an HIA's effectiveness relative to the objectives set out during screening and scoping. In most cases, influencing decisions to protect or promote health is a central objective but by no means the sole outcome of value. As discussed above, additional benefits may include, for example (Wismar et al. 2007; Harris-Roxas and Harris 2011),

- Alerting decision-makers to the more general need to focus on health in future decisions.

- Developing new cross-disciplinary and interagency collaborations.
- Identifying data gaps and questions for future research.
- Establishing a foundation for appropriate monitoring.
- Ensuring that the public has accurate and complete information on adverse and beneficial effects.
- Developing new forecasting methods.
- Improving relationships and collaboration between stakeholders.

Outcome Evaluation

Whereas HIA aims to predict the effects of a decision before it occurs, outcome evaluation assesses whether the implementation of a decision has actual effects on health or health determinants (Parry and Kemm 2005). Outcome evaluation requires a suitable research design, ideally an appropriate comparison group, and data from the monitoring of health outcomes or of changes in health determinants as described above. The committee notes that outcome evaluation considers the effects of the whole decision, including changes made as a result of HIA recommendations. Thus, it is generally not possible to attribute outcomes specifically to HIA recommendations because they are implemented with the decision.

Evaluation of whether a decision has changed specific health outcomes may often be difficult or impossible because of the complex and multifactorial causal pathways involved in many health outcomes, the length of time from implementation of a decision to observable changes in health indicators, and the lack of suitable comparison groups (Quigley and Taylor 2004; Parry and Kemm 2005). However, in some cases, the relationships between the implemented decision and health determinants may be more direct and measurable. Because of the timeframe of proposal implementation and effects on health, outcome evaluation often requires a long-term research commitment. The committee notes that outcome evaluation of policy experiments is a field independent of HIA, and many large-scale social interventions—such as Head Start and Moving to Opportunity—have been subject to outcome evaluation that has included consideration of health or health determinants (Leventhal and Brooks-Gunn 2003; Schweinhart et al. 2005; Frank et al. 2006; Jagannathan et al. 2010). There are, however, no current examples of HIAs in the United States that include outcome evaluation as described here.

Outputs of Monitoring and Evaluation

Monitoring should provide information that allows one to conduct the evaluations noted above. An evaluation plan should have been developed early in the HIA process to guide selection of the appropriate methods for conducting

Elements of a Health Impact Assessment   *81*

evaluations. An evaluation report should be produced at the conclusion of the HIA that includes the following:

- An evaluation of the HIA process against the HIA plan and applicable standards and consideration of whether the process used was appropriate given the decision-making context, needs, objectives, and resources available (a process evaluation).
- A description of the HIA's impact on decision-making (to the extent that salient decisions have occurred by that time) as measured by an accounting of HIA recommendations that were adopted and an evaluation of available evidence that suggests whether and how the HIA played a role in decisions or contributed to changes in decision-makers' knowledge, attitudes, or positions.
- A discussion of whether the HIA achieved its initial objectives.
- Acknowledgement of plans for future outcome evaluation or discussion of limitations that prevent such an evaluation.

Committee Conclusions Regarding Monitoring and Evaluation

Few HIA evaluation data have been published in the United States and relatively few elsewhere. The committee notes that some guides consider evaluation not as a step of HIA but rather as an independent practice that supports the development of the field (see Appendix E). Although completed HIA reports are readily available, peer-reviewed or gray literature that discusses the impacts of specific HIAs is still rare. Evaluation is important for the quality of individual HIAs and for the success of the HIA field as a whole. It is not reasonable to expect decision-makers to adopt HIA widely in the absence of evidence of its effectiveness and value. Consequently, the committee concludes that the lack of attention to evaluation is a barrier that will need to be overcome if HIA practice is to be advanced in the United States.

Evaluation can be thought of in two useful and complementary ways: self-evaluation of the HIA process and impacts and independent external evaluation. Self-evaluation performed by the HIA team—for example, against a set of process objectives or practice criteria—serves quality-assurance aims and can produce insights that will improve the field. Self-evaluation should be considered a valuable step of the HIA process. It may lack the objectivity and rigor of an external evaluation conducted by an experienced evaluator, but it is important because it contributes to a database that informs other efforts in the field and provides basic information about the applications of HIA, the methods and strategies used by HIA practitioners, and the success of and challenges to its use. In contrast, independent evaluation can yield unbiased insights about an HIA from the perspectives of stakeholders and decision-makers, can contribute to a more robust external peer review, and can provide rich information regarding the strengths, weaknesses, and most effective methods and approaches in the field.

The characteristics and approaches of evaluation should be chosen to fit the time, resources, and data available to the HIA team. Building evaluation into the plans for an HIA early in the process may support and reinforce a more deliberate and careful approach to designing and implementing the HIA itself. Although HIA may not always include or provide resources for independent evaluation, more in-depth, independent evaluation will generate more robust conclusions about HIA's effectiveness and best practices in the field and should be given high priority. The committee considers self-evaluation *and* independent evaluation to be essential for moving the field ahead.

Outcome evaluation will continue to be challenging, but it can generate useful information in well-selected cases. Monitoring outcomes can in some cases help to test the validity of predictions and inform future analytic methods. Although there are many potential benefits of undertaking an HIA, one common objective is to inform decisions to promote changes that support improvements in health determinants or health outcomes. For that reason, it is important for the field to define the circumstances under which outcome evaluation may be practicable. Outcome evaluation should be undertaken when available resources and data will allow reasonable judgments regarding the association between the implementation of decisions and observed changes in health outcomes or health determinants.

SUMMARY: WHAT CRITERIA DEFINE A HEALTH IMPACT ASSESSMENT?

This chapter has described HIA categories, defined HIA, discussed current HIA practice, noted variations in practice, and provided the committee's conclusions regarding each step of the HIA process. The discussion recognizes that the practice of HIA varies because it is adapted for use in different decision-making contexts. The variability also reflects a lack of clear criteria that define HIA as a distinct field. On the basis of its review of available literature, HIA guides, and practice standards, the committee has synthesized the key criteria that define HIA and that set it apart from related approaches to public-health practice and policy. Not all HIAs will meet all proposed criteria, but the criteria are intended to describe typical practice. Although deviation from the criteria may occur, a valid and clearly articulated rationale for such deviation should be described when the HIA is reported.

- Health impact assessment is conducted to inform a decision-making process and is intended to be concluded and communicated in advance of the decision that is being assessed.
- It develops the scope of health effects for analysis through systematic consideration of all factors associated with the proposed action that have a potential to influence health, and it narrows the scope to effects that are judged most important for health.

- It identifies a baseline that describes the health status of populations that will be affected by the decision.
- It characterizes health effects according to their nature, direction, intensity, magnitude, distribution, timing and duration, and likelihood.
- It uses the best available evidence to analyze effects on health and health disparities.
- It solicits and responds to input from stakeholders throughout all stages of the process and includes publicly available and accessible documentation of processes, products, and sponsors.
- It recommends measures, in the context of the proposed action, to protect and promote health and reduce health disparities.
- It follows a systematic process that includes screening, scoping, assessment, recommendations, reporting, and monitoring and evaluation.

REFERENCES

AECF (Annie E. Casey Foundation). 2011. About the Annie E. Casey Foundation, Baltimore, Maryland March 2011[online]. Available: http://www.aecf.org/~/media/Pubs/Other/A/AboutCasey/031111_88952_Aboutcasey.pdf [accessed May 13, 2011].

ADA (Atlanta Development Authority). 2005. Atlanta BeltLine Redevelopment Plan. Atlanta Development Authority. November 2005 [online]. Available: http://www.atlantada.com/media/CoverandTableofContents.pdf [accessed Jan. 24, 2011].

Balint, J., P. Boelens, and M. Debello. 2003. Pp. 97-116 in Health Impact Assessment: SEIC (Sakhalin Energy Investment Company) Phase 2 Development. World Health Organization [online]. Available: http://www.who.int/hia/examples/energy/en/HIA_Chps13_18.pdf [accessed July 29, 2011].

Bennear, L.S., and S.M. Olmstead. 2008. The impacts of "right-to-know:" Information disclosure and the violation of drinking water standards. J. Environ. Econ. Manage. 56(2):117-130.

Bhatia, R. 2010. A Guide for Health Impact Assessment. California Department of Public Health. October 2010 [online]. Available: http://www.cdph.ca.gov/pubsforms/Guidelines/Documents/HIA%20Guide%20FINAL%2010-19-10.pdf [accessed Apr. 22, 2011].

Bhatia, R., and E. Seto. 2011. Quantitative estimation in Health Impact Assessment: Opportunities and challenges. Environ. Impact Assess. Rev. 31(3):301-309.

Bhatia, R., and A. Wernham. 2008. Integrating human health into environmental impact assessment: An unrealized opportunity for environmental health and justice. Environ. Health Perspect. 116(8): 991-1000.

Bhatia, R., L. Farhang, M. Gaydos, K. Gilhuly, B. Harris-Roxas, J. Heller, M. Lee, J. McLaughlin, M. Orenstein, E. Seto, L. St. Pierre, A.L. Tamburrini, A. Wernham, and M. Wier. 2009. Practice Standards for Health Impact Assessment (HIA), Version 1. North American HIA Practice Standards Working Group, Oakland, CA. April 2009 [online]. Available: http://www.habitatcorp.com/whats_new/HIA_Practice_Standards_040709_V1.pdf [accessed May 17, 2011].

Bhatia, R., J. Branscomb, L. Farhang, M. Lee, M. Orenstein, and M. Richardson. 2010. Minimum Elements and Practice Standards for Health Impact Assessment (HIA), Version 2. North American HIA Practice Standards Working Group, Oakland, CA.

November 2010 [online]. Available: http://www.sfphes.org/HIA_Tools/HIA_Practice_Standards.pdf [accessed May 23, 2011].

Birley, M. 2005. Health Impact Assessment in multinationals: A case study of the Royal Dutch/Shell Group. Environ. Impact Assess. Rev. 25(7-8):702-713.

BLM (Bureau of Land Management). 2007. Northeast National Petroleum Reserve-Alaska (NPR-A) Draft Supplemental Integrated Activity Plan/Environmental Impact Statement (IAP/EIS). U.S. Department of the Interior, the Bureau of Land Management [online]. Available: http://www.blm.gov/ak/st/en/prog/planning/npra_general/ne_npra/northeast_npr-a_draft.html [accessed Nov. 30, 2010].

BLM (Bureau of Land Management). 2008. Northeast National Petroleum Reserve-Alaska, Supplemental Integrated Activity Plan, Record of Decision. U.S. Department of the Interior, the Bureau of Land Management [online]. Available: http://www.blm.gov/pgdata/etc/medialib/blm/ak/aktest/planning/ne_npra_final_supplement.Par.91580.File.dat/ne_npra_supp_iap_rod2008.pdf. [accessed July 29, 2011].

Castro, A., L. Chen, B. Edison, J. Huang, K. Mitha, M. Orkin, Z. Tejani, D. Tu, L. Wells, and J. Yeh. 2010. Santa Monica Airport Health Impact Assessment (HIA): A Health-Directed Summary of the Issues Facing the Community near the Santa Monica Airport. UCLA Community Health and Advocacy Training Program. February 2010 [online]. Available: http://www.hiaguide.org/sites/default/files/SM_Airport_Health_Impact_Assessment.pdf [accessed May 16, 2011].

Cave, B., and S. Curtis. 2001a. Developing a practical guide to assess the potential health impact of urban regeneration schemes. Promot. Educ. 8(1):12-16.

Cave, B., and S. Curtis. 2001b. Health Impact Assessment for Regeneration Projects, Vols. I and III. Breaking the Cycle, East London and the City Health Action Zone and Queen Mary, University of London, London [online]. Available: http://www.bcahealth.co.uk/links-internal.html [accessed May 16, 2011].

Cave, B., S. Curtis, A. Coutts, and M. Aviles. 2001. Health Impact Assessment for Regeneration Projects. Vol. II. Selected Evidence Base. Breaking the Cycle, East London and the City Health Action Zone and Queen Mary, University of London, London [online]. Available: http://www.bcahealth.co.uk/links-internal.html [accessed May 16, 2011].

Child Health Impact Working Group. 2005. Affordable Housing and Child Health: A Child Health Impact Assessment of the Massachusetts Rental Voucher Program. Child Health Impact Working Group, Boston, MA. June 2005 [online]. Available: http://www.healthimpactproject.org/resources/document/massachusetts-rental-voucher-program.pdf [accessed July 29, 2011].

Cole, B.L., and J.E. Fielding. 2007. Health impact assessment: A tool to help policy makers understand health beyond health care. Annu. Rev. Public Health 28:393-412.

Cole, B.L., M. Wilhelm, P.V. Long, J.E. Fielding, G. Kominski, and H. Morgenstern. 2004. Prospects for health impact assessment in the United States: New and improved environmental impact assessment or something different? J. Health Polit. Policy Law 29(6):1153-1186.

Cole, B.L., R. Shimkhada, J.E. Fielding, G. Kominski, and H. Morgenstern. 2005. Methodologies for realizing the potential of health impact assessment. Am. J. Prev. Med. 28(4):382-389.

Cook, W.K., J. Heller, R. Bhatia, and L. Farhang. 2009. A Health Impact Assessment of the Healthy Families Act of 2009. Human Impact Partners and San Francisco Department of Public Health. June 2009 [online]. Available: http://www.humanimpact.org/component/jdownloads/finish/5/68 [accessed May 16, 2011].

Davenport, C., J. Mathers, and J. Parry. 2006. Use of health impact assessment in incorporating health considerations in decision making. J. Epidemiol. Community Health 60(3):196-201.

DHHS (U.S. Department of Health and Human Services). 2010. Healthy People. Office of Disease Prevention and Health Promotion, U.S. Department of Health and Human Services [online]. Available: http://www.healthypeople.gov/2010/ [accessed May 17, 2011].

EC (European Communities). 2001. Directive 2001/42/EC of the European Parliament and of the Council of 27 June 2001 on the assessment of the effects of certain plans and programmes on the environment. O.J. Eur. Comm. L 197:30-37.

Elliott, E., and S. Francis. 2005. Making effective links to decision-making: Key challenges for health impact assessment. Environ. Impact Assess. Rev. 25(7-8):747-757.

EPA (U.S. Environmental Protection Agency). 1989. Risk Assessment Guidance for Superfund, Vol. 1. Human Health Evaluation Manual Part A. EPA/540/1-89/002. Office of Emergency and Remedial Response, U.S. Environmental Protection Agency, Washington, DC. December 1989 [online]. Available: http://rais.ornl.gov/documents/HHEMA.pdf [accessed June 8, 2011].

EPA (U.S. Environmental Protection Agency). 1992. Framework for Ecological Risk Assessment. EPA/63-R-92/001. Risk Assessment Forum, U.S. Environmental Protection Agency, Washington, DC. February 1992.

EPA (U.S. Environmental Protection Agency). 2009. Red Dog Mine Extension Aqqaluk Project. Final Supplemental Environmental Impact Statement. Prepared for U.S. Environmental Protection Agency, Seattle, WA, by Tetra Tech, Inc., Anchorage, AK. October 2009 [online]. Available: http://www.reddogseis.com/Docs/Final/Front_Matter.pdf [accessed Nov. 30, 2010].

Fehr, R. 1999. Environmental health impact assessment: Evaluation of a 10 step model. Epidemiology 10(5):618-625.

Frank, D.A., N.B. Neault, A. Skalicky, J.T. Cook, J.D. Wilson, S. Levenson, A.F. Meyers, T. Heeren, D.B. Cutts, P.H. Casey, M.M. Black, and C. Berkowitz. 2006. Heat or eat: The Low Income Home Energy Assistance Program and nutritional and health risks among children less than 3 years of age. Pediatrics 118(5):e1293-e1302.

Fredsgaard, M.W., B. Cave, and A. Bond. 2009. A Review Package for Health Impact Assessment Reports of Development Projects. Leeds, UK: Ben Cave Associates Ltd [online]. Available: http://www.bcahealth.co.uk/pdf/hia_review_package.pdf .

Halifax Initiative Coalition. 2006. One Step Forward, One Step Back: An Analysis of the IFC's Sustainability Policy, Performance Standards and Disclosure. Ottawa, Canada: Halifax Initiative Coalition [online]. Available: http://www.ifc.org/ifcext/policyreview.nsf/AttachmentsByTitle/HalifaxReport/$FILE/IFC-Analysis-HI-Final.pdf [accessed Feb. 3, 2011].

Harris, P., B. Harris-Roxas, E. Harris, and L. Kemp. 2007. Health Impact Assessment: A Practical Guide. Sidney, Australia: Centre for Health Equity Training, Research and Evaluation, the University of New South Wales. August 2007 [online]. Available: http://www.hiaconnect.edu.au/files/Health_Impact_Assessment_A_Practical_Guide.pdf [accessed May 9, 2011].

Harris-Roxas, B., and E. Harris. 2011. Differing forms, differing purposes: A typology of health impact assessment. Environ. Impact Assess. Rev. 31(4):396-403.

Heller, J., J. Lucky, and W.K. Cook. 2009. Crossings at 29th St. / San Pedro St. Area Health Impact Assessment. Human Impact Partners, Oakland, California [online].

Available: http://www.hiaguide.org/sites/default/files/Crossings_LA_29thSt_HIA_FullReport.pdf [accessed May 17, 2011].

Human Impact Partners. 2009a. Newsroom: Paid Sick Days HIAs. Human Impact Partners [online]. Available: http://www.humanimpact.org/press [accessed May 16, 2011].

Human Impact Partners. 2009b. Past Projects: Paid Sick Days Legislation. Human Impact Partners [online]. Available: http://www.humanimpact.org/past-projects [accessed May 16, 2011].

Human Impact Partners. 2010. Past Projects: Vehicle Miles Traveled Legislation. Human Impact Partners [online]. Available: http://www.humanimpact.org/past-projects [accessed May 24, 2011].

ICMM (International Council on Mining and Metals). 2010. Good Practice Guidance on Health Impact Assessment. London, UK: International Council on Mining and Metals [online]. Available: http://www.icmm.com/page/35457/good-practice-guidance-on-health-impact-assessment [accessed May 16, 2011].

IFC (International Finance Corporation). 2006. Policy on Disclosure of Information. International Finance Corporation. April 30, 2006 [online]. Available: http://www.ifc.org/ifcext/enviro.nsf/AttachmentsByTitle/pol_Disclosure2006/$FILE/Disclosure2006.pdf [accessed May 17, 2011].

IFC (International Finance Corporation). 2007. Guidance Note 4. Community Health, Safety and Security. International Finance Corporation. July 31, 2007 [online]. Available: http://www.ifc.org/ifcext/sustainability.nsf/AttachmentsByTitle/pol_GuidanceNote2007_4/$FILE/2007+Updated+Guidance+Note_4.pdf [accessed Feb. 3, 2011].

IFC (International Finance Corporation). 2010. International Finance Corporation Performance Standard 4 – Rev -0.1. Community Health, Safety, and Security. International Finance Corporation. April 14, 2010 [online]. Available: http://www.ifc.org/ifcext/policyreview.nsf/AttachmentsByTitle/Phase2_PS4_English_clean/$FILE/CODE_Progress+Report_AnnexB_PS4_Clean.pdf [accessed Feb. 3, 2011].

IPIECA/OGP (International Petroleum Industry Environmental Conservation Association and International Association of Oil and Gas Producers). 2005. A Guide to Health Impact Assessments in the Oil and Gas Industry. International Petroleum Industry Environmental Conservation Association, and International Association of Oil and Gas Producers [online]. Available: http://www.hiaconnect.edu.au/files/HIA_in_OG.pdf [accessed May 17, 2011].

Jagannathan, R., M.J. Camasso, and U. Sambamoorthi. 2010. Experimental evidence of welfare reform impact on clinical anxiety and depression levels among poor women. Soc. Sci. Med. 71(1):152-160.

Johnson, B.L. 1999. The role of adaptive management as an operational approach for resource management agencies. Conserv. Ecol. 3(2):8 [online]. Available: http://www.ecologyandsociety.org/vol3/iss2/art8/ [accessed May 24, 2011].

Kemm, J. 2001. Health impact assessment: A tool for healthy public policy. Health Promot. Int. 16(1): 79-85.

Kuo, T., C.J. Jarosz, P. Simon, and J.E. Fielding. 2009. Menu labeling as a potential strategy for combating the obesity epidemic: A health impact assessment. Am. J. Public Health 99(9):1680-1686.

Lester, C., S. Hayes, S. Griffiths, G. Lowe, and S. Hopkins. 1999. Implementing a strategy to address health inequalities: A health authority approach. Public Health Med. 1(3):90-93.

Leventhal, T., and J. Brooks-Gunn. 2003.Moving to opportunity: An experimental study of neighborhood effects on mental health. Am J Public Health 93(9):1576-1582.

McAuliffe, M. 2009. Developers of Proposed Springfield Biomass Plant tell Public Health Council: "Nothing Less Than the Best". The Republican, December 2, 2009. MassLive.com [online]. Available: http://www.masslive.com/news/index.ssf/2009/12/developers_of_proposed_springf.html [accessed May 19, 2011].

McHugh, S, S. Maruca, J. Lilien, and A. Manning. 2006. Environmental, Social and Health Impact Assessment (ESHIA) Process. Paper No. 98224-MS. SPE International Health, Safety & Environment Conference, 2-4 April 2006, Abu Dhabi, UAE. Society of Petroleum Engineers.

Mindell, J., E. Ison, and M. Joffe. 2003. A glossary for health impact assessment. J. Epidemiol. Community Health 57(9):647-651.

Mindell, J., A. Boaz, M. Joffe, S. Curtis, and M. Birley. 2004. Enhancing the evidence base for health impact assessment. J. Epidemiol. Community Health 58(7):546-551.

Mindell, J.S., A. Boltong, and I. Forde. 2008. A review of health impact assessment frameworks. Public Health 122(11):1177-1187.

Mindell, J., J. Biddulph, L. Taylor, K. Lock, A. Boaz, M. Joffe, and S. Curtis. 2010. Improving the use of evidence in health impact assessment. Bull. World Health Organ. 88(7):543-550.

Miranda, M.L., M.H. Keating, and S.E. Edwards. 2008. Environmental justice implications of reduced reporting requirements of the Toxics Release Inventory Burden Reduction Rule. Environ. Sci. Technol. 42(15):5407-5414.

Morgan, R.K. 2011. Health and impact assessment: Are we seeing closer integration? Environ. Impact Assess. Rev. 31(4):404-411.

Neidell, M. 2009. Information, avoidance behavior, and health: The effect of ozone on asthma hospitalizations. J. Human Res. 44(2):450-478.

NRC (National Research Council). 1983. Risk Assessment in the Federal Government: Managing the Process. Washington, DC: National Academy Press.

NRC (National Research Council). 1996. Understanding Risk: Informing Decisions in a Democratic Society. Washington, DC: National Academy Press.

NRC (National Research Council). 2009. Science and Decisions: Advancing Risk Assessment. Washington, DC: National Academies Press.

Parry, J., and A. Stevens. 2001. Prospective health impact assessment: Pitfalls, problems, and possible ways forward. BMJ 323(7322):1177-1182.

Parry, J.M., and J.R. Kemm. 2005. Criteria for use in the evaluation of health impact assessments. Public Health 119(12):1122-1129.

Petticrew, M., S. Cummins, L. Sparks, and A. Findlay. 2006. Validating health impact assessment: Prediction is difficult (especially about the future). Environ. Impact Assess. Rev. 27(1):101-107.

PHAC (Public Health Advisory Committee). 2005. A Guide to Health Impact Assessment: A Policy Tool for New Zealand, 2nd Ed. Wellington, New Zealand: PHAC [online]. Available: http://www.phac.health.govt.nz/moh.nsf/pagescm/764/$File/guidetohia.pdf [accessed May 9, 2011].

Presidential/Congressional Commission on Risk Assessment and Risk Management. 1997. Framework for Environmental Health Risk Management – Final Report, Vol. 1. [online]. Available: http://www.riskworld.com/nreports/1997/risk-rpt/pdf/EPAJAN.PDF [accessed July 28, 2011].

Quigley, R.J., and L.C. Taylor. 2004. Evaluating health impact assessment. Public Health. 118(8):544-552.

Quigley, R., L. den Broeder, P. Furu, A. Bond, B. Cave, and R. Bos. 2006. Health Impact Assessment: International Best Practice Principles. Special Publication Series No. 5. Fargo: International Association for Impact Assessment. September 2006 [online]. Available: http://www.iaia.org/publicdocuments/special-publications/SP5.pdf [accessed May 6, 2011].

Roscam Abbing, E.W. 2004. HIA and national policy in the Netherlands. Pp. 177-189 in Health impact Assessment: Concepts, Theory, Techniques and Applications, J. Kemm, J. Parry, and S. Palmer, eds. Oxford: Oxford University Press.

Ross, C.L. 2007. Atlanta Beltline: Health Impact Assessment. Center for Quality Growth and Regional Development, Georgia Institite of Technology, Atlanta, GA [online]. Available: http://www.healthimpactproject.org/resources/document/Atlanta-Beltline.pdf [accessed May 18, 2011].

Schweinhart, L.J., J. Montie, Z. Xiang, W.S. Barnett, C.R. Belfield, and M. Nores. 2005. Lifetime Effects: The High/Scope Perry Preschool Study through Age 40. Ypsilanti, MI: High/Scope Press.

Scott-Samuel, A. 1996. Health impact assessment. BMJ 313(7051):183-184.

SFCC (Federation of Swedish County Councils). 1998. Focusing on Health: How Can the Health Impact of Policy Decisions be Assessed? Federation of Swedish County Councils, Stockholm, Sweden [online]. Available: http://www.lf.se/hkb/engelskversion/general.htm.

SFDPH (San Francisco Department of Public Health). 2011. Assessing the Health Impacts of Road Pricing Policy Proposals. Program on Health Equity and Sustainability, San Francisco Department of Public Health [online]. Available: http://www.sfphes.org/HIA_Road_Pricing.htm [accessed July 8, 2011].

Steinemann, A. 2000. Rethinking human health impact assessment. Environ. Impact Assess. Rev. 20(6):627-645.

Thomson, H. 2008. HIA forecast: Cloudy with sunny spells later? Eur. J. Public Health 18(5):436-438.

UPH (Upstream Public Health). 2009. Health Impact Assessment on Policies Reducing Vehicle Miles Traveled in Oregon Metropolitan Areas. Upstream Public Health [online]. Available: http://www.upstreampublichealth.org/publications?type=reports [accessed June 13, 2011].

Vaccaro, A., and P. Madsen. 2009. Corporate dynamic transparency: The new ICT-driven ethics? Ethics Inform. Technol. 11(2):113-122.

Veerman, J., J. Barendregt, and J. Mackenback. 2005. Quantitative health impact assessment: Current practice and future directions. J. Epidemiol. Community Health. 59(5):361-370.

Wernham, A. 2007. Inupiat health and proposed Alaskan oil development: Results of the first Integrated Health Impact Assessment/Environmental Impact Statement of proposed oil development on Alaska's North Slope. EcoHealth 4(4):500-513.

Wernham, A. 2011. Health impact assessments are needed in decision making about environmental and land-use policy. Health Aff. 30(5):947-956.

WHO (World Health Organization). 1999. Health Impact Assessment: Main Concepts and Suggested Approaches-the Gothenburg Consensus Paper. Brussels: European Centre for Health Policy, WHO Regional Office for Europe.

Wismar, M., J. Blau, K. Ernst, and J. Figueras. 2007. The effectiveness of Health Impact Assessment: Scope and Limitations of Supporting Decision-Making in Europe.

Copenhagen: World Health Organization [online]. Available: http://www.euro.who.int/__data/assets/pdf_file/0003/98283/E90794.pdf [accessed May 18, 2011].

Wolf, S. 1996. Fear and loathing about the public right to know: The surprising success of the emergency planning and community Right-to-Know Act. J. Land Use Environ. Law 11(2)217-325.

World Bank. 2010. The World Bank Policy on Access to Information. Document No. 54873. World Bank. July 1, 2010 [online]. Available: http://www-wds.worldbank.org/external/default/WDSContentServer/WDSP/IB/2010/06/03/000112742_20100603084843/Rendered/PDF/548730Access0I1y0Statement01Final1.pdf [accessed May 18, 2011].

4

Current Issues and Challenges in the Development and Practice of Health Impact Assessment

Chapter 2 discussed the need for health-informed decisions and the advantages of using health impact assessment (HIA) to evaluate the potential health consequences of an array of projects, plans, programs, and policies. Chapter 3 provided a framework for HIA and highlighted critical elements of each step in the HIA process. This chapter identifies and explores several topics considered by the committee to be the most salient issues or challenges for the successful emergence, development, and practice of HIA. First, the committee addresses how health should be defined for HIA and how its definition influences the application and scope of HIA practice. Types of decisions that are potential candidates for HIA are then considered. The committee next reviews several methodologic issues for HIA, including the need to balance timely information with variable data quality, expectations for quantitative estimates, synthesizing conclusions on dissimilar health effects, assigning monetary values to health outcomes, enabling stakeholder participation, and the benefits of a peer-review process for HIA. The committee then examines the potential for conflicts of interest among HIA practitioners, sponsors, and funders and considers whether it is realistic to expect the practice of HIA to result in a change in the decision being made. The committee concludes with a discussion of how HIA is related to the consideration of human health effects in environmental impact assessment (EIA) as required by the National Environmental Policy Act (NEPA) and similar state laws.

DEFINING HEALTH FOR HEALTH IMPACT ASSESSMENT

How health is defined and considered by society and government institutions—that is, what is or is not considered by practitioners, decision-makers, and stakeholders to have relevance to and a bearing on health—ultimately establishes the boundaries for HIA practice. That determination will clearly influence which decisions are considered appropriate subjects for HIA and which health effects are considered to be within its scope. Many have recognized that a narrow definition of health or factors that influence health probably limits the scope, application, and value of the practice.

The constitution of the World Health Organization (WHO) considers health broadly and states that "health is a state of complete physical, mental, and social well-being and not merely the absence of disease or infirmity" (WHO 1946, p. 100). Although there are many definitions of health—many less expansive than the WHO definition—there is a growing consensus that health at the individual and population levels is shaped by a combination of genetic, behavioral, social, economic, political, and environmental factors. As discussed in Chapter 2, the root causes or determinants of health include the quality and accessibility of infrastructure, such as housing, schools, parks, and transportation systems; the safety of the environment and economic security; the number and quality of social interactions; cultural characteristics, such as diet; and the level of equity and social inclusion. It is therefore essential that those many determinants be considered in defining the boundaries of HIAs. In the present committee's view, HIA must be concerned broadly with individual and public health and all its social, cultural, political, economic, and environmental determinants.

Using such a broad definition of health has clear implications for which decisions may be subject to HIA, the scope of issues and measures used to characterize health in HIA, and how health effects are weighed in relation to competing outcomes. In general, the public-health practice has traditionally defined health more narrowly and focused on disease, morbidity, and longevity. Thus, many decisions that affect health determinants have been considered outside the scope and mandate of public-health institutions. As discussed in Chapter 2, the failure to attend to the broader health determinants—for example, economic conditions—have contributed to avoidable disease and health disparities (CSDH 2008). However, broadening the definition of health has implications for the work of other sectors and their relationships with each other and with public health. Expecting institutions outside the health-care and public-health sectors to advance public-health interests will be challenging because actions needed to protect and promote health are often in conflict with the interests and objectives of other sectors. Critics may question whether addressing public-health objectives should be weighed more heavily than meeting the objectives of the sector in whose domain a decision is being debated. Ultimately, broadening the definition of health creates the setting where tradeoffs among health and other social objectives can be made transparently. Recent calls for public agencies to consider and take actions to improve health indicate changing attitudes and the need to create a more multidisciplinary approach to public health (CSDH 2008). The committee supports the recent government actions and emphasizes the need to define health broadly in the practice of HIA but recognizes that implementation will require some care to balance health with the many other considerations that are important to any given decision.

ARE ALL DECISIONS POTENTIAL CANDIDATES FOR HEALTH IMPACT ASSESSMENT?

A frequent question—given the breadth of potential applications of HIA—is whether there is a limit on the types of decisions to which the practice might be applied. For example, is HIA better suited to decisions in particular policy sectors (such as education, urban planning, and finance), to a particular scale (such as policy vs project) or jurisdictional level, or to particular health outcomes? The question is important because there are few formal requirements for analyzing the health effects of decisions except for the requirements for health analysis under NEPA and state environmental policy acts (SEPAs), and as demand for HIA grows, there will be a greater need to target its applications efficiently.

The broad definition of health discussed above suggests that a wide array of decisions—including some of those made in almost all government sectors on local, state, national, and international scales—may be appropriate candidates for HIA (Harris-Roxas and Harris 2011). A review of the sectors in which HIAs have been completed in the European Union (EU) (Wismar et al. 2007) and in the United States (Dannenberg et al. 2008; HIA-CLIC 2010; RWJF/PEW 2011) underscores this breadth of potential applications (see Table 4-1). Although most U.S. examples reflect applications in the transportation, housing, or urban-planning sectors, there is growing diversity in the United States and a wider diversity in the existing spectrum of EU applications. The growth may be because of greater experience with and public support for HIA and increased public recognition of the many determinants of health.

In the committee's view, restricting the spectrum of HIA practice to particular decisions, sectors, decision scales, jurisdictional levels, or health issues is unwarranted. At this early stage, there is no evidence to suggest that HIA is more important, appropriate, or effective in any particular decision context. On the contrary, HIA may be useful across a broad array of decision contexts, including many decision types to which it has not yet been applied. Furthermore, new global health challenges are likely to emerge from issues related to atmospheric and climate change, population growth, food and land scarcity, revolutionary industrial technologies (such as nanotechnology and gene modification), globalization, and economic inequities (WWF/ZSL/Global Footprint Network 2010). For example, a changing climate and an increase in extreme weather events will have many effects, including widespread effects on health (Costello et al. 2009; Luber and Prudent 2009). Public policy in general and public health and HIA in particular must recognize the emerging challenges and support the identification of adaptive and preventive strategies. HIA may play a substantive role in emphasizing the importance of the emerging issues to public health and to policy-makers and stakeholders.

TABLE 4-1 Health Impact Assessment by Sector

Sector[a]	European Union[b]	United States[c]
Transport	27 (17%)	21 (28%)
Housing or urban planning	23 (15%)	28 (38%)
Environment	18 (11%)	3 (4%)
Health	14 (9%)	0 (0%)
Employment	10 (6%)	4 (5%)
Social care	8 (5%)	0 (0%)
Finance	8 (5%)	0 (0%)
Energy	7 (4%)	8 (11%)
Agriculture	7 (4%)	2 (3%)
Industry	4 (3%)	5 (7%)
Education	3 (2%)	3 (4%)
Tourism	2 (1%)	0 (0%)
Multiple sectors	17 (11%)	0 (0%)
Other	10 (6%)	0 (0%)
Total	158	74

[a]The list of sectors was taken from Wismar et al. (2007). The authors did not provide the criteria used to determine whether a report was considered an HIA, and they did not explicitly define how HIAs were categorized into sectors. There is clearly potential for policies, plans, programs, and projects to fall into two or more categories.
[b]Wismar et al. (2007) was used as the source for the EU data.
[c]HIAs conducted in the United States were identified from lists maintained by the Health Impact Project (RWJF/PEW 2011), the University of California, Los Angeles (HIA-CLIC 2010), and Dannenberg et al. (2008) and from committee experience. To be included in the table, an HIA must have been identified as such by the authors and must have documented at least some steps of the HIA process. The committee recognizes that the list may not be up to date or exhaustive, but the table shows examples of the sectors that do HIAs. As seen here, many more HIAs have been carried out in the EU than in the United States. In both the United States and the EU, HIAs are carried out most often in the transportation, housing, and urban-planning sectors.

Although most decisions have the potential to affect health, conducting HIAs of all decisions is clearly not practical or expected. Accordingly, HIA proponents should try to select decisions that have the greatest opportunities for advancing public-health goals and promoting the awareness of the health implications of decision-making. As described in Chapter 3, one purpose of the screening step in HIA is to focus HIA on high-priority topics by explicitly considering the value of conducting HIA in a particular situation. For example, findings of an HIA of a proposed decision may be appropriately applied to a similar decision in another context or on another scale (for example, regulations for labor standards on city, state, and federal scales), so the value of conducting another HIA may be diminished.

The committee emphasizes that as long as HIA is conducted as a voluntary process, it will be difficult to ensure that it is directed at the most important health priorities and decision opportunities. Aside from the limited analysis of health effects that is currently conducted within the regulatory structure of EIA, practitioners of or those funding or sponsoring HIA are in most cases selecting decisions by using ad hoc mechanisms based on their own interests and goals and are considering a limited set of candidate decisions (for example, land-use projects in a particular locality). Without clear mandates, screening criteria, and procedural rules for HIA, the selective approach to conducting an HIA may miss decisions for which HIA would have value and produce some HIAs that have little utility for decision-makers or stakeholders.[1] Furthermore, HIA could conceivably contribute to health inequities if more socioeconomically or politically advantaged communities develop greater capacity to demand HIA or if health issues that are highlighted in HIA are focused on the health needs of the advantaged.

In contrast, institutional rules for HIA—for example, rules articulated in laws at the local, state, or federal level—could establish consistent procedures for the field and ensure that a sufficiently broad set of candidate decisions are screened. For example, decisions subject to HIA might be selected and ranked on the basis of the likelihood of addressing the Healthy People 2020 objectives of the U.S. Department of Health and Human Services or on the basis of the most realistic opportunities to address environmental injustice or to reduce health inequities. Institutional rules could effectively narrow a large number of candidate decisions to a manageable ordered set, enhance the use of HIA, and advance its rationale and equitable use. Such rules could also help to organize and direct the creation of a coherent and systematic body of knowledge about decision-related health effects and analytic methods that could be used for HIA.

---

[1]The committee notes that although screening is considered an essential step in the HIA process, there is little published documentation or evaluation on the implementation of the screening step and thus little information on cases in which HIA might have been considered and not conducted. Some but not all HIA reports explain the rationale for conducting the assessment, but still there is little understanding of why HIAs have or have not been pursued.

Thus, the committee finds that any future policies, standards, or regulations for HIA should include explicit criteria for identifying and screening candidate decisions and rules for providing oversight for the HIA process; such criteria and rules would promote the utility, validity, and sustainability of HIA practice.

BALANCING THE NEED TO PROVIDE TIMELY, VALID INFORMATION WITH THE REALITIES OF VARIED DATA QUALITY

A substantial challenge facing HIA practitioners is the quality and availability of evidence on which to base predictions about health effects (Mindell et al. 2004; Petticrew 2007; Veerman et al. 2007; Bhatia and Seto 2011; Mindell et al. 2010). More broadly, scholars acknowledge that the studies and empirical evidence linking improvements in health directly to changes in specific public policies are sparse (Curtis et al. 2002; Dow et al. 2010; Graham 2010). Furthermore, many decisions occurring outside the health sector have not previously been seen as important for health, so they have typically not been the subject of rigorous empirical health research. Thus, making prospective judgments about the effects of policy decisions on health is challenging, and concerns about validity arise in the face of variable and often sparse evidence.

The committee emphasizes that concerns about validity must be balanced against the reality that decisions that are not informed by health analysis have the potential to harm health (see Chapter 2); this implies that the degree to which the evidence limits judgments must always be weighed against the potential severity and scope of harm that could occur if available information were not considered. But practitioners should also consider the risk, both to optimal decision-making and to the legitimacy of the field, that is inherent in overstating the precision or certainty of health-effect estimates provided by HIA.

There are challenges to addressing concerns about the validity of HIA predictions. Regardless of evidence-related constraints, HIA must operate in the context of practical realities and timelines of the decision-making process, and HIA reports need to express clearly the quality of evidence and the degree of confidence in inferences drawn from the evidence. The committee notes that society regularly accepts such practical limitations in making policy decisions and that predictive certainty or causal certainty would be an impractical standard for HIA.

Practical and agreed-on methods for addressing concerns about validity are needed, and the committee offers three strategies, discussed below, that should help to improve the validity of health-effects judgments made in the context of variable evidence:

- Consider diverse evidence sources by using expertise in multiple disciplines.
- Assess the quality of available evidence.

- Include a strategy for assessing and managing uncertainty.

HIA practitioners can also learn from the health-risk-assessment field where some analysts have demonstrated the ability to adapt their analyses to varied evidence, ranging from data extrapolated from the literature when local information is lacking to primary reliance on local data that leverage knowledge and statistical power from the broader literature (Hubbell et al. 2009).

Consider Diverse Evidence Sources by Using Expertise in Multiple Disciplines

As discussed in Chapter 3, many types of evidence can be used in HIA, including peer-reviewed academic studies; unpublished, publicly available studies and databases, that is, gray literature; survey, monitoring, or interview data specific to the affected population or to the policy, plan, program, or project in question; the experience of people who will be affected by the proposed changes; and expert opinion. The committee recommends that practitioners review all the available evidence systematically (Mindell et al. 2010). Practitioners should use published, peer-reviewed systematic reviews—such as those conducted by the Cochrane Collaboration, WHO, the U.S. Environmental Protection Agency, and other authoritative bodies—if they are available. Although studies conducted within the population that might be affected are ideal, they may not exist or be feasible to conduct, so analysis will turn to literature and data on other populations.

HIA is necessarily a multidisciplinary practice. It is often, although not exclusively, carried out by public-health professionals, but it almost always requires access to experts in the core domains that are affected by the proposal under consideration. Multiple disciplines will help to reveal differences of opinion, will provide the team with access to a variety of evidence and analytic methods, and may provide a more robust critique of methods, findings, and conclusions. The participation of multiple public agencies—such as health, planning, and transportation agencies—not only will contribute expertise but may ensure that the process addresses questions pertinent to the decision at hand and thus increase the likelihood that the recommendations are actionable and will be adopted. Furthermore, conducting HIA as a multidisciplinary practice can assist in developing ownership and commitment for health goals among multiple institutional and disciplinary sectors. The committee nevertheless recognizes that it can be challenging to conduct multidisciplinary analysis or to manage the participation of multiple agencies or participants.

In selecting evidence and evaluating quality, practitioners should recognize their own biases and the biases of decision-makers, project proponents, or HIA sponsors. Biases may affect the value attached to particular types of evidence. For example, evidence from consultation, which can be more readily dismissed as hearsay or anecdotal, may not be accorded as much weight as the quantitative modeling of environmental exposure or economic effects (Ozonoff 1994). Other biases related to conflicts of interest are discussed later in this chapter.

Evaluate Evidence Quality

HIA practitioners should select the strongest evidence and analytic methods that are available for a particular decision context. For transparency, it is equally important to state the rationale for choosing particular evidence or methods when alternatives are available. Key factors that should be considered in determining whether to use a given study or dataset include the relationship between study end points and the issues evaluated in the HIA, the quality of the data and their statistical power, the adequate assessment of factors that could impede causal inference (that is, the internal validity of an empirical study) (Susser 1986; Rothman and Greenland 1998; Weed 2005), and the applicability of the evidence to the target population (that is, external validity). The quality of evidence used in HIA may also be assessed according to the core standards of the discipline in which the data originate; for example, epidemiologic studies should generally be evaluated according to quality standards for epidemiologic studies with attention to such issues as the potential for bias and confounding.

There are no uniform standards for evaluating all potential evidence that might be used in HIA given the diversity of applications and of the evidence base. However, many of Bradford Hill's (Hill 1965) causal criteria—such as strength of association, consistency of evidence among studies and data sources, coherence with known facts of the exposure and disease, and analogy to similar situations—could be applied to HIA when evaluating the likelihood of health effects. Other criteria could be developed to extrapolate findings on study populations to the target populations for specific decisions. And criteria could be developed in ways that are specific to the needs of different policy contexts.

Setting any uniform evidence standards carries some risk of limiting the scope of health effects and pathways assessed in HIA. In health risk assessment, even with assumptions and acceptance of uncertainty, evidence requirements in practice have constrained analysis to a limited set of exposures and outcomes (NRC 2009). Even if HIA practice evolves standardized approaches for the analysis of particular decisions, determinants, or health effects, there will be a need for flexibility to address new and emerging issues.

Characterize and Manage Uncertainty

Uncertainty will always be present, and impact assessments—including HIAs—should characterize and manage the uncertainty to the extent possible and practicable that is inherent in the analyses and decisions. Although uncertainty should not be ignored in HIA, it should also not paralyze the decision process. Furthermore, there may be situations in which the magnitude of uncertainty is large enough to make selection among competing alternatives challenging, but the potential impact may be important enough to justify intervention in the face of that uncertainty.

Managing uncertainty in HIA can include planning how the analysis will address uncertainties and establishing procedures to characterize or reduce key uncertainties. Uncertainty in the analysis of health effects can be characterized in a variety of ways, ranging from qualitative descriptions to quantitative analysis. Quantitative analyses of uncertainty are common in related fields, such as health risk assessment, and are relevant if the key health effects are quantified in the HIA. Distributions of estimates based on various assumptions have been presented in HIA (Schram-Bijkerk et al. 2009), but it typically includes only a subset of assumptions for which distributions can be readily quantified and omits some major sources of uncertainty for which quantification is impractical. More generally, although formal propagation and quantification of uncertainty can be helpful in elucidating the influence of key assumptions, they contribute to a lack of methodologic transparency for many stakeholders and raise potential issues with timeliness. At a minimum, uncertainties in assumptions used to support health-effect characterization should be described qualitatively. In other words, HIA practitioners should evaluate and document the uncertainty of their conclusions by describing the evidence on which their conclusions are based and by identifying any limitations, gaps, or weaknesses in the assumptions. That exercise should go beyond parametric uncertainty described in individual studies to consider broader questions, such as whether a measure of exposure used in HIA was a reliable proxy for personal exposure or whether an exposure-response function extracted from the literature can be generalized to the population of interest.

Similar issues have been confronted in the domain of health risk assessment. The National Research Council report *Science and Decisions: Advancing Risk Assessment* (NRC 2009) concluded that the plans for uncertainty analysis should be discussed during the scoping process to ensure that the information generated meets the needs of decision-makers and to avoid unnecessarily complex (and untimely) uncertainty analyses. For example, if a mitigation or alternative is readily available and affordable for managing a health impact of concern and has greater benefits than other alternatives, a formal treatment of uncertainty may be unnecessary. Planning of how uncertainty will be evaluated and managed—including quantitative and qualitative elements—should be a component of the scoping process of HIA (see Chapter 3). The plan should consider how stakeholders may wish to see uncertainty information presented, including the method of presentation and the emphasis on distributions vs expected values vs upper or lower bound values for aspects that can be quantified. Various approaches for characterizing the sophistication of uncertainty analysis (Pate-Cornell 1996; IPCS 2006) could be adapted for HIA, as could previously recommended strategies for addressing and communicating uncertainty in complex multifactorial models (NRC 2007) and in cost-effectiveness analysis (Briggs 2000; Claxton 2008).

As suggested, characterization of uncertainty will often need to go beyond quantitative methods to include other forms of information. Using a deliberative group process to arrive at judgments is a nonquantitative way to manage uncertainty and to moderate the effects of individual and organizational values and biases. HIA conducted as a deliberative group process that involves open discussion and debate among the stakeholders may also be useful in generating judgments that will be widely accepted. The National Institutes of Health uses a deliberative process to achieve consensus on many clinical issues in medicine (NIH 2011); this approach may have value in managing uncertainties in HIA.

As the practice of HIA evolves, there may be uncertainties and data limitations that call for a set of practical assumptions to avoid subjective or ad hoc variations in analyses. In the practice of health risk assessment, default science-policy assumptions are used to allow the analysis to proceed with incomplete information (NRC 1983). For example, carcinogens whose mode of action is unknown are generally assumed to have linear dose-response functions at low doses, whereas nonlinear dose-response functions are assumed for noncancer health effects in the absence of specific evidence about mode of action or chemical pharmacodynamics. Similarly, when exposure or dose information is lacking, numerous default assumptions are used to capture breathing rates, drinking-water consumption, and various other behaviors or activities. In each case, there is inadequate information on the specific pollutants and settings of interest, but the analysis proceeds with assumptions derived from a combination of evidence from analogous situations and science-policy judgments. Although the variety of applications of HIA makes specific default science-policy assumptions difficult to formalize, the concept can be used to provide more transparency and interpretability. Over time, for policy contexts in which numerous HIAs are conducted, default science-policy assumptions could be generated and could facilitate comparability among HIAs. Regardless, HIA practitioners should explicitly describe where key judgments or assumptions were made, whether or not uncertainty can be formally characterized, and what implications the assumptions have on the HIA recommendations. In that way, the ultimate choice among competing options can be made by decision-makers given their preferences regarding action in the face of uncertainty.

BENEFITS AND CHALLENGES OF QUANTITATIVE ESTIMATION

Some decision-makers and HIA users expect HIA to provide quantitative estimates of health effects. Quantitative estimates of health effects have a number of desirable properties: they provide an indication of the magnitude of health effects, they can be easily compared with existing numerical criteria or thresholds that define the significance of particular effects, they allow one to make more direct comparisons among alternatives, and they provide inputs for economic valuation (see section "Assigning Monetary Values to Health Consequences" below). They can be produced when there has been sufficient empirical research on relationships between particular determinants and health outcomes. Accordingly, quantification is most feasible if a causal relationship can be inferred and if there is an externally valid effect measure or a defined exposure-response relationship (Hertz-Picciotto 1995; Fehr 1999; Mindell et al. 2001; O'Connell and Hurley 2009; Bhatia and Seto 2011). If information is lacking or uncertain, quantification may still be possible with the use of assumptions and inferences based on information drawn from analogous situations. In other situations, such as when assumptions are not defensible, quantitative estimates should not be advanced.

HIAs have applied quantitative techniques to decisions to estimate health effects related to expected changes in infectious-disease risks, traffic hazards, environmental pollutants, housing conditions, and tobacco and alcohol consumption (see Box 3-4; Veerman et al. 2005; O'Connell and Hurley 2009; Bhatia and Seto 2011). For example, quantitative impact-assessment methods have been used to estimate human health externalities associated with different fuels in Europe; the analysis was used to inform member states about the impacts of various fuels for electricity (such as nuclear fuel, coal, and natural gas) and was considered in numerous policy analyses, including strategies for internalizing external costs and development of sustainable-transport policies (O'Connell and Hurley 2009). Noise exposures and particulate-matter concentrations were modeled and associated with sleep disturbance and premature mortality, respectively, in local-scale assessments of residential development in San Francisco and waterfront development in Oakland; the assessments used quantitative exposure-response functions from the epidemiologic literature (Bhatia and Seto 2011). The effects of rezoning in San Francisco on pedestrian injuries were modeled on the basis of multivariate regression models derived from geocoded accident data and site characteristics, and changes in body-mass index and accident risks associated with increased walking of children to school in Sacramento were estimated by using data derived from epidemiologic investigations (Bhatia and Seto 2011).

Regardless of the advantages, relying exclusively on quantitative estimation in HIA presents some drawbacks. First, quantification has high information requirements. Given the breadth of health effects potentially considered in HIA, the sparse data available to support quantitative approaches, and the variability in practitioner capacity, it would be challenging if not impossible to expect all HIAs to predict all potentially important health effects quantitatively. Thus, an HIA that presents only quantifiable results would present only a partial accounting of health effects if not all important effects are amenable to quantification (Veerman et al. 2005; O'Connell and Hurley 2009). Second, because quantification can be resource-intensive, it may require more time than allowed for the evaluation of a policy, plan, program, or project. Third, a quantitative approach has implications for communicating the process and results to a wider audience because the methods are typically highly technical and include assumptions that may be difficult to communicate outside the technical team. Quantitative estimates may create an unwarranted impression of objectivity, precision, and importance and lead a reader to place importance or credence in quantified results even if assumptions and measures used in the analysis are based on subjective choices (O'Connell and Hurley 2009). Stakeholders, including lay audiences, may lose trust in the process, especially if they suspect that assumptions in calculations are influenced by the biases of those conducting or sponsoring the assessments (Ozonoff 1994; NRC 2009; O'Connell and Hurley 2009).

Overall, quantitative estimates of health effects have value and should be provided when the data and resources allow and when they are responsive to decision-makers' and stakeholders' information needs. This statement, however, should not imply exclusion of health effects from the analysis for which causal linkages have been made but quantification is impractical. Part of the scoping phase of HIA discussed in Chapter 3 should involve explicit consideration of which exposures and outcomes, if any, would be amenable to quantification and whether such analysis is feasible within the decision timeframe. To manage some of the challenges related to communication outlined above, the technical procedures and assumptions in quantitative analysis should be articulated clearly and explicitly. Approaches to characterize quantitative or technical information and communicate it to decision-makers have been described in detail elsewhere (see, for example, NRC 1989, 1996).

CHARACTERIZING MULTIPLE HEALTH EFFECTS

An HIA analyzes and reports findings on multiple health effects, so providing a simple conclusion is challenging. For example, an HIA conducted on a decision of whether to build a new rail line might evaluate its effects on sleep, asthma symptoms, and traffic injuries. In most cases, those health effects will be described with different units and measures and thus cannot be summarized by using the same unit of measurement; that is, it is not possible simply to add findings expressed in different health metrics. An important challenge is to synthesize and present results on dissimilar health effects in a manner that is intelligible and useful to stakeholders and decision-makers.

The most common approach in HIA is to describe and characterize each effect separately (see Chapter 3) and allow users to make judgments about the cumulative nature of the effects. The committee endorses that approach even if a summary measure of effects is used. Generally, decision-makers must balance multiple desirable and adverse effects related to a decision and will need to "weight" or assign values to them on the basis of institutional rules, constituent preferences, or some other approach. Keeping effects separate and assigning values allow decision-makers to consider tradeoffs among health and nonhealth effects clearly. As described in *Improving Risk Communication*, "reducing different kinds of hazard to a common metric (such as number of fatalities per year) and presenting comparisons only on that metric have great potential to produce misunderstanding and conflict and to engender mistrust of expertise" (NRC 1989, p. 52). The committee emphasizes the importance of characterizing adverse and beneficial effects separately in considering health disparities that could result from a decision; the distributional effects could be hidden or disappear if all effects are combined into one measure.[2]

As indicated above, an alternative way to present findings is to use a summary measure to translate estimated effects on disparate health end points into a single comparable unit, such as quality-adjusted life years (QALYs) (Hammit 2002), disability-adjusted life years, and healthy-years equivalent. Such health utility measures allow for disparate health outcomes to be weighted and combined, and they can include outcomes that are important for public health but are often omitted or underemphasized in health risk assessments (for example, mental illness). The health utility measures, however, bring assumptions that need to be recognized; for example, QALYs focus on years of remaining life expectancy and thus place greater weight on the life and well-being of a child than that of an elderly person. The committee recommends the consideration and application of summary measures in contexts where quantification is possible and the outcomes are amenable to assignment of quality weights or disability weights. However, as stated above, each health outcome should also be individually reported, and multiple summary measures should be used, when it is practical, to determine whether decisions are robust to the weighting scheme and to societal preferences among outcomes and populations.

ASSIGNING MONETARY VALUES TO HEALTH CONSEQUENCES

The health consequences of a decision can be characterized according to their economic or monetary valuation. Although monetary effects clearly are not health effects themselves, many decision-makers and stakeholders may give substantial consideration to the economic value of effects, and economic valuation of health effects can facilitate comparison with the costs and benefits of competing alternatives (Brodin and Hodge 2008).

Economic valuation has several constraints and is not appropriate in all circumstances. First, the wide array of end points may not be amenable to monetary valuation. Second, monetary valuation of health outcomes has implicit and explicit weighting of outcomes and populations that may or may not reflect the values and priorities of decision-makers. For example, willingness to pay will tend to be greater among populations that have greater wealth and will tend to be lower among those who are facing competing risks (Hammitt 2002). Third, some populations may bear a disproportionate share of the health costs of a decision, and others a disproportionate share of the health gain. Those distributional effects can be hidden in cost-benefit analysis conducted at a societal level

---

[2]The committee notes that distributional effects can be evaluated descriptively or quantitatively, and available statistical techniques enable relationships among impact inequalities and socioeconomic or demographic factors to be examined quantitatively (Kakwani et al. 1997; Mitchell 2005).

but would be potentially valuable information for those who incur the costs and for those who receive the benefits. Fourth, monetary valuation of health outcomes can pose a substantial communication challenge for affected parties and other stakeholders and may distract from the findings of an HIA. In spite of those caveats, monetary valuation of health outcomes may be a useful approach in some decision contexts, such as those in which alternative decision choices might require implementing economically costly mitigations.

If economic analysis is conducted as part of HIA, it is important to maintain the distinction between HIA, which provides judgments of health effects, and cost-benefit analysis, which provides a more comprehensive analysis of all economic benefits and costs of a decision. Economic valuation of health effects is common in existing cost-benefit analyses of federal regulations; however, HIA should not be characterized as or confused with cost-benefit analysis.

VALUING AND ENABLING STAKEHOLDER PARTICIPATION

Chapter 3 emphasizes the importance of stakeholder engagement and participation in HIA and echoes the guidance provided repeatedly in the context of environmental risk assessment and risk management (PCCRARM 1997; NRC 2009). Individuals and organizations that are not part of the technical assessment team have the potential to make valuable contributions at each stage of the HIA process. Information gained through stakeholder involvement helps to illuminate important issues and focus the scope of an HIA on the most important or contested issues (Corburn and Bhatia 2007; Farhang et al. 2008; Corburn 2009). It can improve the quality and specificity of an analysis by, for example, highlighting local living conditions, prevalent health issues, and potential effects that might not be visible to practitioners from outside the community (Elliot and Williams 2004; Parry and Kemm 2005). Stakeholder involvement contributes to a more democratic planning or decision-making process by providing a structured and effective way for knowledge to be exchanged among those involved in planning and designing a proposal, those responsible for a decision, and those likely to be affected by the decision. It also helps to ensure that various stakeholder concerns receive adequate attention and that HIA recommendations are realistic and practicable.

The importance of including different perspectives and worldviews is highlighted by the experience of indigenous people whose perspectives and ways of thinking have often challenged knowledge used in, values underpinning, and processes for decision-making. The environment is of paramount importance to indigenous communities because many rely heavily on the land and natural resources for their subsistence, including their socioeconomic, cultural, spiritual, and physical survival (Kwiatkowski et al. 2009). For many indigenous groups, the "term *environment* does not distinguish between humanity and everything else; humans are part of the environment as much as the fish, wildlife, air, and trees" are (Kwiatkowski 2011, p. 447). Furthermore, the timeframe for

EIA is typically shorter than that used when elders assess issues that face their communities (Williams 2010). For example, the constitution of the Iroquois Nations stipulates a period of seven generations over which to consider implications of any actions (Murphy 2001; Haudenosaunee Confederacy 2010). The knowledge and worldviews of indigenous people provide important insights that would not be known to people outside the community and illustrate why it is so important to provide opportunities for local input and influence and not to assume that all groups have a similar perspective.

Ensuring that stakeholders, including the public, are able to participate effectively in HIA is described in Chapter 3 as an essential element of practice (WHO 1999; Parry and Kemm 2005; IFC 2006; Quigley et al. 2006; Fredsgaard et al. 2009; Bhatia et al. 2010). But how or indeed whether practitioners enable stakeholders to participate in HIA varies widely (Kearney 2004; Mindell et al. 2004; Mahoney et al. 2007; Dannenberg et al. 2008). The variation may be attributable to the time and resources available for the HIA, to how high a priority HIA practitioners or sponsors give to participation, to a concern that participation may interfere with or impede progress toward the sponsors' objectives, or to differences in the type and scale of the decision to which the HIA is to be applied (for example, local vs national level). However, it must be recognized that achieving representative participation is challenging, requires experience and particular skills, and may take different forms.

The decision context and the objectives of an HIA will influence who should be engaged, the challenges and opportunities for engaging key stakeholders, and the final selection of specific approaches to engage various stakeholders. For example, project-related decisions that will have direct and immediate implications for local neighborhoods should engage stakeholders from those communities. In contrast, national legislative decisions are more likely to involve representatives of interest-based or constituency-based organizations or possibly elected officials of constituencies that will be affected by legislation. Going beyond broad representative participation may not be necessary or feasible for an HIA of a national policy.

Techniques for stakeholder engagement and involvement are many and varied and can be chosen to suit a specific decision but need to address the barriers and challenges identified for each stakeholder group. Although open community meetings are likely to lend themselves to projects at a local level, other techniques (such as focus groups) can be adapted for any level by ensuring that they include key stakeholder communities and organizations that represent the groups most likely to be affected. Other approaches include interactive Web-based communications that facilitate effective exchanges among practitioners, sponsors, stakeholders, and the public and provide opportunities for stakeholders to review and comment on scope, data sources, findings, and recommendations (UNECE/REC 2007). Stakeholder engagement strategies that solicit and respond to comments on HIA reports only after they have been completed are restricted in their ability to take into account stakeholder concerns in the analysis and are typically viewed as reactive by stakeholders and the public. Whenever it is possible, strategies for stakeholder participation should extend beyond that minimal standard.

Formal oversight or advisory groups can be effective for continuing involvement, such as steering committees that are comprised of practitioners and stakeholders and provide oversight or direction and technical advisory committees that extend the expertise or range of disciplines brought to the HIA (Corburn and Bhatia 2007; Farhang et al. 2008). Formal collaboration agreements can also be used to define the roles and expectations of practitioners and stakeholders. In creating and working with groups, efforts must be made to ensure that members are representative, that differences in technical knowledge or power do not exclude members from full participation, and that disagreements among members are managed effectively. Conducting an HIA on local and regional policies, programs, or projects with the assistance of local community-based organizations that have deep local knowledge and networks is an effective way to achieve involvement of community members that have historically been excluded from decision-making (Wier et al. 2009).

Effective stakeholder participation potentially faces a number of challenges; as noted earlier, these are likely to depend on the scale of the decision for which HIA is being conducted. Participatory processes can favor those who have more resources and expertise and exclude local community or lay stakeholders. For example, groups that have fewer social and economic resources may be the least likely to participate. An equitable HIA process depends on strong efforts to identify and minimize barriers to participation and to ensure adequate representation for those unable to participate directly, for example, through elected officials in the case of national decisions. For HIAs of local and regional decisions, factors that can inhibit or prevent participation by individuals or groups that may be affected by the decisions vary—for example, structural issues, such as limited collective organization or lack of trust in public processes; poor access to elected decision-makers; and practical considerations, such as language or literacy barriers and the requirement to manage competing life needs. Similarly, for HIAs of national decisions (in which stakeholders may include constituency or interest groups or elected officials), people who are economically, socially, or linguistically marginalized may encounter particular challenges to full participation or representation. Efforts to address the challenges can include various strategies and again depend on the scale of the decision. For local or regional decisions, engagement of diverse stakeholders may include hosting meetings at venues in the community, providing translation and child care, scheduling interactions around work demands and important cultural events, and identification of formal and informal leaders in the community for continuing participation. For national decisions, efforts to ensure engagement of a broad array of stakeholders may include identification of regional or national interest groups that represent those likely to be affected by the decision and elected representatives from districts or regions likely to be affected by a decision. External facilitation of stakeholder engagement and involvement may be an effective option in some cases.

THE BENEFITS OF A PEER-REVIEW PROCESS FOR HEALTH IMPACT ASSESSMENT

An important quality-control mechanism in the research process is peer review of the research plan or strategy and of the report that describes the analysis and results. Independent peer review provides a measure of credibility and legitimacy of findings and is commonly used in applied scientific disciplines to monitor practitioner conformity with established practices. HIA is different from primary scientific research in that it involves the application and interpretation of evidence in a particular decision context. Although premises underlying HIA judgments are often based on peer-reviewed evidence, several additional aspects of the HIA process might benefit from peer review.

HIA involves the selective identification of issues and the selective use of evidence. Peer review might identify overlooked issues or indicate opportunities to improve data or methods. Judgments about health effects are inferences based on evidence and observations that use reasoning and assumptions. Many of the procedural aspects of HIA—such as selection of evidence and transparency in the reporting stage of HIA—are instrumental in the acceptability and utility of findings in the decision process and may benefit from review by HIA experts who are independent of the process. Thus, peer review might increase the legitimacy of conclusions and their acceptance and utility in the decision-making process.

Regardless of the potential benefits, an accepted peer-review process for HIA would need to overcome several challenges. There are many stages at which peer review could theoretically be applied, and the multidisciplinary nature of HIA requires varied expertise and raises the issue of which people or teams would be best suited to conduct such a review. The involvement of teams of reviewers at multiple steps in the process could substantially increase the time and effort required to complete an HIA and could therefore make it less practical for decisions that need to be made in the short-term and for HIA teams with few resources. Peer review would also require agreed-on criteria, and at present there are no uniformly accepted criteria with which to judge the quality of an HIA. Furthermore, given the need for a flexible and adaptive tool that is applicable to an array of decision contexts, flexibility of quality criteria may be needed. In addition, peer review would need to be distinguished from public comment, and a process would need to be created to demonstrate responsiveness to peer-review comments.

Currently, peer review appears to be undertaken only intermittently, and the committee notes that the benefits of peer review need to be weighed carefully against the risk of delays that would render the HIA less relevant to the decision that it is intended to inform and the added costs and time that could restrict the use of HIA in some cases. Given the potential benefits, however, a formal peer-review process could be used at least in targeted large-scale, high-profile cases in which the benefits of added scrutiny and rigor would outweigh the disadvantages of added delay and process. In other cases, practitioners could consider a less structured process. For example, practitioners could request informally that colleagues review their HIAs or that they review particularly challenging, complex, or controversial aspects of their findings. It is common for HIA practitioners to get advice from other practitioners during the course of an HIA, and some implement technical advisory committees. Those approaches might achieve some of the objectives of an independent formal peer review. Development of accepted standards, databases, models, and default assumptions in the field would enable HIAs to be peer-reviewed with a consistent approach (Fredsgaard et al. 2009; Bhatia et al. 2009, 2010).

MINIMIZING CONFLICTS OF INTEREST OF SPONSORS AND PRACTITIONERS OF HEALTH IMPACT ASSESSMENT

Impact assessments, including HIA, are conducted on decision proposals that are often contested among polarized and disparate interests and stakeholders. Regulatory assessment practices have been criticized as selectively representing interests, particularly those of development-project proponents. Given the decision-driven nature of HIA, even when there are substantial resources and high-quality data, results of HIA may still be contested and be subject to accusations of bias (Milner et al. 2003). Ensuring that the process by which HIA is conducted and the conclusions and recommendations that are produced at the end of the process are impartial, credible, and scientifically valid is paramount to the effectiveness of the practice (Veerman et al. 2006). To the extent feasible, those who conduct the assessment should strive to avoid real and perceived conflicts of interests.

The source of HIA funding is a common challenge to objectivity in HIA. Bias toward a funder's interests is a well-recognized problem in many other forms of analysis and assessment. For example, Lexchin et al. (2003) found that results were more likely to favor pharmaceutical companies when they sponsored studies than when others sponsored them. In the practice of EIA, in which assessments of economic-development proposals are commonly funded and conducted by development proponents, assessors may feel substantial pressure to hide or minimize adverse effects of the proposals or to emphasize favorable effects (Morgan 1998). That bias can be reflected in issues and alternatives evaluated, methods used, assumptions made, results presented, and mitigations offered. An HIA funded by a development proponent may be similarly vulnerable to influence and may lead to a process that is more likely to find a result consistent with the interests of the sponsor.

Private commercial interests are not the only entities that may exert influence on HIA via funding or sponsorship. Private grant-makers (such as philanthropies) currently provide a substantial share of the funding for HIA that is conducted voluntarily in the United States. Philanthropies may influence the process or findings of HIA in several ways, for example, by directing HIA funding to assess specific health issues (such as air pollution or obesity), and this could potentially bias the scope of the assessment and the associated results. Mission-driven grant-makers may have strong expectations that HIA will produce substantive change in the issues and interests that they champion, or they may wish to see clear evidence that the HIA influenced a decision.

Government agencies sponsoring HIAs may also have interests that exert influence on HIA practices or conclusions. For example, government agencies may be less welcoming of results that potentially raise criticisms of their actions, identify their oversights, or challenge their positions. Like development-project proponents, public agencies may have a preferred decision outcome and may be interested in ensuring that HIA reflects favorably on that alternative. Such conflicts may be heightened if the agency conducting the HIA is also the responsible decision-maker.

In some cases, stakeholders or practitioners may decide to champion, sponsor, or conduct an HIA because of a strong interest in a specific decision outcome. They may seek to use HIA as a means to support or advocate for a particular policy outcome (Harris-Roxas and Harris 2011). In such cases, there may be a substantial risk of introducing bias into the HIA process.

The committee emphasizes that a lack of trust by any stakeholders in the HIA practitioner can undermine the legitimacy and influence of HIA. Therefore, it is important to guard against and mitigate the conflicts of interest described above. It may be useful for future practice guidance to establish a clear line between a practitioner's role in conducting HIA and later efforts toward advocacy of particular decision outcomes.[3] Although public entities may be somewhat less vulnerable to influence because of public funding sources, oversight mechanisms, and requirements for transparency, they are not immune to influence. Public-health agencies that have the necessary experience and expertise and the confidence of stakeholders may be in good positions to conduct or coordinate HIA given their mandate to protect public health. Other mechanisms to manage or mitigate influence may include the eventual creation of a dedicated public funding source to conduct HIA and a process of independent peer review of HIA as discussed above.

MANAGING EXPECTATIONS: INFORMATION MAY NOT CHANGE DECISIONS

HIA clearly is intended to inform decisions, but information alone does not necessarily change decisions. The committee recognizes that the underlying motivation of HIA is to make policy and decisions that are more cognizant of and aligned with the interests of public health. Informing decision-makers can certainly influence attitudes and preferences and lead to more responsive health-

---

[3]The committee distinguishes between advocacy (that is, trying to influence the decision outcome) and explaining or educating decision-makers on the findings and recommendations made in an HIA.

supporting actions, but it is not reasonable to base the effectiveness of HIA on whether it changes decisions in which health is only one of many considerations and over which the HIA team lacks decision-making authority. Furthermore, support and legitimacy of the practice may be compromised if an HIA is conducted explicitly as a mechanism for decision advocacy.

It is reasonable to hope that identifying valid information about the health-related harms or benefits of a decision will motivate decision-makers to take protective actions.[4] However, generating high-quality health information and effectively communicating it does not ensure that the information is given high priority in the decision-making process or triggers action. HIA is not designed or practiced as a mechanism to regulate decision-making directly (that is, to require responsive actions if impacts exceed criteria). Although effective communication can raise awareness of and attention to health concerns, improved knowledge alone cannot necessarily change the ideology, interests, and attitudes of decision-makers. Health is typically one of many objectives under consideration in a given policy question, and decision-makers and other stakeholders are reasonably influenced by factors and tradeoffs beyond the quality or findings of an HIA.

Although HIA does not guarantee particular decision outcomes, providing publicly available information on health effects clearly is a mechanism of influence. Thomas Jefferson famously stated that "information is the currency of democracy." In the case of EIA under NEPA, the purpose of the process was to give environmental consequences due consideration (Yost 2003). Although the courts observed that NEPA provides protection only from the harm of uninformed decision-making, not from adverse environmental consequences themselves, informing decisions has substantial power. Institutional rules for EIA have opened decision-making to public scrutiny, raised the profile of environmental considerations, and altered the norms and practices of public and private organizations in a way that is more protective of the environment (Canter and Clark 1997; Cashmore et al. 2007).

Given that HIA practitioners are not typically in decision-making positions, effective and broad communication of findings is essential to the informational objectives and to later influence on decision-making. Communication of findings may be optimal when there is an opportunity for assessors to have discussion with decision-makers and stakeholders. Effective dissemination may require educating decision-makers about the public-health evidence underlying conclusions and may require consideration of and response to criticisms about the findings or about the efficacy or feasibility of recommendations. Many others, apart from practitioners, may be in strong positions to communicate findings of HIA and their importance in the decision-making process. The accounting of health effects by HIA should allow the public and stakeholders to use informa-

---

[4]The committee notes that revisions might be made in a proposal or its alternatives in anticipation of an HIA being conducted; such changes might not ultimately be considered to be a result of the HIA but might not have occurred if the HIA were not planned.

tion in the political process to advance health interests. The political use of HIA evidence—like other types of information disclosed to the public—should be viewed as a normal mechanism of its influence on decisions.

ADVANCING REQUIREMENTS FOR HEALTH ANALYSIS IN ENVIRONMENTAL IMPACT ASSESSMENT

This chapter has thus far discussed HIA as it is practiced outside the context of EIA. NEPA and some SEPAs explicitly require the identification and analysis of health effects when EIA is conducted, and there are various views on how HIA might be related to or support health-effects analysis in the EIA process (see Appendixes A and F for further discussion). Although the scope of health-effects analysis has been limited in U.S. EIA practice, some argue that greater use should be made of NEPA and related state laws as a mechanism for health-informed policy-making given that it has the same substantive ends as HIA (Bhatia and Wernham 2008; Wernham 2009; Morgan 2011). Others, however, contend that EIA has become too rigid a practice to accommodate the attention and resources needed for conducting a comprehensive analysis of health effects (Cole et al. 2004) and that attention should be focused on the independent practice of HIA.

The committee is keenly aware of the time and resources that NEPA compliance can entail. However, assessment of direct, indirect, and cumulative health effects in EIA under NEPA and many SEPAs is a matter of law, not discretion, when it is likely to add important information that is relevant to decision-making (see Appendix A for further discussion). Therefore, when legal requirements call for an integrated analysis of health effects in the EIA process, this analysis should be conducted in observance of the same procedures and standards as for any other environmental or social effects being considered. In the case of health, those procedures would arguably mirror the general steps of HIA as described in Chapter 3 and would include a description of the baseline health status of the population; an analysis of the direct, indirect, and cumulative health consequences of the proposed action and alternatives; and a consideration of potential mitigation measures to address the health concerns identified by the analysis. If adequately conducted, the steps would be consistent with and might be considered equivalent to conducting an HIA.

To date, however, despite the requirements for the analysis of health effects in EIA, the consideration of health effects in EIA practice has been limited, and public-health experts have rarely been involved in the EIA process (Davies and Sadler 1997; Steinemann 2000; Hilding-Rydevik et al. 2006). The limited practice may partly reflect that historically, NEPA practice has been shaped primarily by pressure and litigation brought by environmental groups, and public-health advocates have only rarely demanded health-effects analysis. The limited practice also may reflect the resource constraints facing many public-health departments and more generally the lack of familiarity with EIA practice. Challenges to changing EIA practice to include more substantive health analysis include resistance on the part of agencies leading EIA to invest time and resources in routine health analysis, lack of familiarity with or expertise in public health on the part of agencies that commonly lead EIAs, and limited relationships with local, state, and tribal health authorities or others that have the capacity to conduct public-health analyses (Cole et al. 2004; Hilding-Rydevik et al. 2006; Corburn and Bhatia 2007; Bhatia and Wernham 2008). Because of those challenges, some HIA practitioners have voiced concern that, in contrast with independent HIA, integrating health into EIA might produce a narrow consideration of health effects (Cole et al. 2004). Furthermore, it is possible that agencies responsible for EIA may give less importance to health effects than to other environmental concerns; consider only health effects that are quantifiable with traditional methods, such as human health risk assessment; or allocate insufficient funding for health-effects analysis. Those concerns are valid, but the committee notes that the problem is not unique to the setting of integrated EIA. Currently, HIA conducted independently of EIA has no mechanism to monitor or ensure the adequacy of resources and breadth of analysis, and like EIA, the scope of HIA has been limited by practitioner decisions and available resources (Dannenberg et al. 2008).

Considering those important challenges, the committee concludes that improving the integration of health into EIA under NEPA and related state laws is needed and would serve the mission of public health and the goals of HIA. Federal agencies file thousands of EIA documents each year. Decision-making that is subject to EIA requirements at the federal and state levels includes a wide array of projects, programs, and policies that have broad importance for health. Furthermore, health issues are among the most common concerns raised by affected communities.

Agencies formally responsible for conducting EIAs and practitioners in the field of public health have an interest in improving the consideration of health in EIA. When health effects are relevant to a proposed action, agencies responsible for conducting EIA should seek out appropriate public-health expertise and should invite tribal, federal, state, or local health agencies to participate as cooperating agencies (40 C.F.R Sections 1501.6, 15018.5). Adequate resources should be accorded to health-effects analysis in EIA. Similarly, public-health officials need to take a more active role in EIA by offering appropriate information and expertise to aid the analysis.

Recent experience in the field has demonstrated that a greater consideration of direct, indirect, and cumulative health effects can be accomplished in EIA if the associations described are well supported by public-health theory and evidence (Wernham 2009; Bhatia and Wernham 2008; Morgan 2011). The official submission of findings by public-health agencies into the public record (for example, via public comment on draft environmental documents) has triggered comment and analysis by responsible agencies (Bhatia 2007). Interagency partnerships that have involved public health during an EIA have reduced skepticism on the part of agencies unfamiliar with public health and HIA, have fostered a broader shared understanding of potential health effects, and have led to health-protective mitigations and alternatives to the proposals that were assessed (Wernham 2007; Bhatia and Wernham 2008; Morgan 2011). In several cases under both NEPA and the California Environmental Quality Act, the scope of health effects and alternatives considered has been substantially augmented with the financial resources and expertise needed to conduct related analyses. In other cases, health-effects analyses have trigged substantive mitigations. There remain, however, substantial opportunities to improve the consideration and analysis of health effects under NEPA and SEPAs. Conflicts and negotiation of interests among environmental assessors and health professionals concerning values, objectives, scope, and use of information should be expected in the course of developing a stronger integrated practice (Morgan 2011).

Anecdotally, some concerns have been raised that broadening the scope of health analysis in NEPA may increase the potential for litigation. The committee finds little factual support for that view; indeed, only rarely has EIA litigation been based on inadequacy of health analysis. Indeed, the failure to address potentially important effects and substantive concerns is a leading reason for litigation under NEPA and may result in an order to the agency to address the omissions; this could cause delays in projects. Given that there is increasing attention to the relationship between public policies and health in the United States, the failure to address potentially important health effects may leave agencies more vulnerable to litigation; ensuring a comprehensive analysis of health in EIA may be a good way for agencies to avoid such risks.

CONCLUSIONS

- Although there are many definitions of health, there is a growing consensus that health at the individual and population levels is shaped by a combination of genetic, behavioral, social, economic, and environmental factors. It is essential that those determinants be considered in defining the boundaries of HIA.
- It is not necessary or appropriate to conduct HIA for all decisions at the local, state, or federal levels; however, restricting the spectrum of HIA practice to particular decision types, institutional sectors, decision scales (for example, policy, program, or plan), or jurisdictional levels or to specific health issues is not warranted. The use of HIA should be focused on applications in which there is the greatest opportunity to protect or promote health and to raise awareness of the health consequences of proposed decisions.
- The committee finds that three strategies should help to improve the validity of health-effects predictions made in the context of varied evidence: consider diverse evidence sources by using expertise in multiple disciplines, assess the quality of available evidence, and implement a strategy for assessing and managing uncertainty.

Current Issues and Challenges in Development and Practice of HIA   *113*

- Quantitative estimates of health effects in HIA have a number of desirable properties, but it is impractical to expect quantitative estimates in all applications of HIA given the sparseness of quantitative data on associations between many policy decisions and health.
- An HIA that analyzes multiple dissimilar health effects should describe and characterize each effect separately and consider ways to provide aggregate or summary measures of dissimilar effects.
- Although HIA is not a cost-benefit analysis, economic valuation of health effects may be requested by decision-makers and should be considered when relevant data are available. As with any HIA component, economic valuation should be provided with a discussion of key assumptions and methodologic limitations.
- The committee emphasizes the importance of stakeholder engagement and participation in HIA. Information gained through stakeholder involvement can help to focus the scope of the HIA and improve its quality and specificity by highlighting local living conditions, prevalent health issues, and potential effects that might not be visible to practitioners outside the community.
- A formal peer-review process for HIA could increase the acceptability or utility of conclusions on health effects or related mitigations and should be considered when the benefits of added scrutiny and rigor would outweigh the disadvantages of added delay and process.
- To the extent feasible, practitioners conducting HIA should strive to avoid real and perceived conflicts of interest. It may be useful for future practice guidance to establish a clear line between the practitioner's role in conducting HIA and later advocacy of particular decision outcomes. A dedicated public funding source and a process of independent peer review of HIA may help in managing or mitigating conflicts of interest.
- HIA aims to influence attitudes and preferences and leads to actions that support health, but HIA may not change decisions when health is only one of many considerations. Conducting HIA as a mechanism for advocacy may compromise support for and legitimacy of the practice.
- Improving the integration of health into EIA practice under NEPA and related state laws would serve the mission of public health and the purpose of HIA and EIA. Despite known challenges, agencies responsible for EIA and public-health practitioners share responsibility for improving the consideration of health effects in EIA practice. However, to ensure reasonable priority of health issues under NEPA, public-health agencies should be afforded a substantive role in the scoping and oversight of health-effects analysis in EIA, and health-effects analysis must be afforded resources commensurate with the task.
- The committee concludes that any future policies, standards, or regulations for HIA should include explicit criteria for identifying and screening candidate decisions and rules for providing oversight for the HIA process; such criteria and rules would promote the utility, validity, and sustainability of HIA practice.

REFERENCES

Bhatia, R. 2007. Protecting health using environmental impact assessment. Am. J. Public Health 97(3):406-413.

Bhatia, R., and E. Seto. 2011. Quantitative estimation in Health Impact Assessment: Opportunities and challenges. Environ. Impact Assess. Rev. 31(3):301-309.

Bhatia, R., and A. Wernham. 2008. Integrating human health into environmental impact assessment: An unrealized opportunity for environmental health and justice. Environ. Health Perspect. 116(8): 991-1000.

Bhatia, R., L. Farhang, M. Gaydos, K. Gilhuly, B. Harris-Roxas, J. Heller, M. Lee, J. McLaughlin, M. Orenstein, E. Seto, L. St. Pierre, A.L. Tamburrini, A. Wernham, and M. Wier. 2009. Practice Standards for Health Impact Assessment (HIA), Version 1. North American HIA Practice Standards Working Group, Oakland, CA. April 2009 [online]. Available: http://www.habitatcorp.com/whats_new/HIA_Practice_Standards_040709_V1.pdf [accessed May 17, 2011].

Bhatia, R., J. Branscomb, L. Farhang, M. Lee, M. Orenstein, and M. Richardson. 2010. Minimum Elements and Practice Standards for Health Impact Assessment (HIA), Version 2. North American HIA Practice Standards Working Group, Oakland, CA. November 2010 [online]. Available: http://www.sfphes.org/HIA_Tools/HIA_Practice_Standards.pdf [accessed May 23, 2011].

Briggs, A.H. 2000. Handling uncertainty in cost-effectiveness models. Pharmacoecon. 17(5):479-500.

Brodin, H., and S. Hodge. 2008. A Guide to Quantitative Methods in Health Impact Assessment. Swedish National Institute of Public Health, Östersund, Sweden. [online]. Available: http://www.fhi.se/PageFiles/6057/R2008-41-Quantitative-Methods-in-HIA.pdf [accessed May 20, 2011].

Canter, L., and R. Clark. 1997. NEPA effectiveness: A survey of academics. Environ. Impact Assess. Rev. 17(5):313-327.

Cashmore, M., A. Bond, and D. Cobb. 2007. The contribution of environmental assessment to sustainable development: Toward a richer empirical understanding. Environ. Manage. 40(3):516-530.

Claxton, K. 2008. Exploring uncertainty in cost-effectiveness analysis. Pharmacoecon. 26(9):781-798.

Cole, B.L., M. Wilhelm, P.V. Long, J.E. Fielding, G. Kominski, and H. Morgenstern. 2004. Prospects for health impact assessment in the United States: New and improved environmental impact assessment or something different? J. Health Polit. Policy Law 29(6):1153-1186.

Corburn, J. 2009. Toward the Healthy City: People, Places and the Politics of Urban Planning. Cambridge: The MIT Press.

Corburn, J., and R. Bhatia. 2007. Health Impact Assessment in San Francisco: Incorporating the social determinants of health into environmental planning. J. Environ. Plann. Manage. 50(3):323-341.

Costello, A., M. Abbas, A. Allen, S. Ball, S. Bell, R. Bellamy, S. Friel, N. Groce, A. Johnson, M. Kett, M. Lee, C. Levy, M. Maslin, D. McCoy, B. McGuire, H. Montgomery, D. Napier, C. Pagel, J. Patel, J.A. Puppim de Oliveira, N. Redclift, H. Rees, D. Rogger, J. Scott, J. Stephenson, J. Twigg, J. Wolff, and C. Patterson. 2009. Managing the health effects of climate change. Lancet 373(9676):1693-1733.

CSDH (Commission on Social Determinants of Health). 2008. Closing the Gap in a Generation: Health Equity through Action on the Social Determinants of Health. Geneva: World Health Organization [online]. Available: http://www.who.int/social_determinants/thecommission/finalreport/en/index.html [accessed May 11, 2011].

Curtis, S., B. Cave, and A. Coutts. 2002. Is urban regeneration good for health? Perceptions and theories of the health impacts of urban change. Environ. Plann. C 20(4):517-534.

Dannenberg, A.L., R. Bhatia, B.L. Cole, S.K. Heaton, J.D. Feldman, and C.D. Rutt. 2008. Use of health impact assessment in the U.S.: 27 case studies, 1999-2007. Am. J. Prev. Med. 34(3):241-256.

Davies, K., and B. Sadler. 1997. Environmental Assessment and Human Health: Perspectives, Approaches, and Future Directions. Ottawa: Health Canada [online]. Available: http://dsp-psd.pwgsc.gc.ca/Collection/H46-3-7-1997E.pdf [accessed May 20, 2011].

Dow, W.H., R.F. Schoeni, N.E. Adler, and J. Stewart. 2010. Evaluating the evidence base: Policies and interventions to address socioeconomic status gradients in health. Ann. N.Y. Acad. Sci. 1186:240-251.

Elliott, E., and G. Williams. 2004. Developing a civic intelligence: Local involvement in HIA. Environ. Impact Assess. Rev. 24(2):231-243.

Farhang, L., R. Bhatia, C.C. Scully, J. Corburn, M. Gaydos, and S. Malekfzali. 2008. Creating tools for healthy development: Case study of San Francisco's Eastern Neighborhoods Community Health Impact Assessment. J. Public Health Manag. Pract. 14(3):255-265.

Fehr, R. 1999. Environmental health impact assessment: Evaluation of a 10 step model. Epidemiology 10(5):618-625.

Fredsgaard, M.W., B. Cave, and A. Bond. 2009. A Review Package for Health Impact Assessment Reports of Development Projects. Leeds, UK: Ben Cave Associates Ltd [online]. Available: http://www.bcahealth.co.uk/pdf/hia_review_package.pdf.

Graham, H. 2010. Where is the future in public health? Milbank Q. 88(2):149-168.

Hammitt, J.K. 2002. QALYs versus WTP. Risk Anal. 22(5):985-1001.

Harris-Roxas, B., and E. Harris. 2011. Differing forms, differing purposes: A typology of health impact assessment. Environ. Impact Assess. Rev. 31(4):396-403.

Haudenosaunee Confederacy. 2010. The Great Law of Peace. The Haudenosaunee Confederacy [online]. Available: http://www.haudenosauneeconfederacy.ca/greatlawofpeace.html [accessed Nov. 27, 2010].

Hertz-Picciotto, I. 1995. Epidemiology and quantitative risk assessment: A bridge from science to policy. Am. J. Public Health 85(4):484-491.

HIA-CLIC (Health Impact Assessment Clearinghouse Learning and Information Center). 2010. Completed HIAs. University of California, Los Angeles [online]. Available: http://www.hiaguide.org/hias [accessed Oct. 26, 2010].

Hilding-Rydevik, T., S. Vohra, A. Ruotsalainen, A. Pettersson, N. Pearce, C. Breeze, M. Hrncarova, Z. Liekovska, K. Paluchova, L. Thomas, and J. Kemm. 2006. Health Aspects in EIA. European Commission Sixth Framework Program [online]. Available: http://www.umweltbundesamt.at/fileadmin/site/umweltthemen/UVP_SUP_EMAS/IMP/IMP3-Health_Aspects_in_EIA.pdf [accessed May 23, 2011].

Hill, A.B. 1965. The environment and disease: Association or causation? Proc. R. Soc. Med. 58:295-300.

Hubbell, B.J., N. Fann, and J.I. Levy. 2009. Methodological considerations in developing local-scale health impact assessments: Balancing national, regional and local data. Air Qual. Atmos. Health 2(2):99-110.

IFC (International Finance Corporation). 2006. Performance Standard 1: Social and Environmental Assessment and Management Systems. Pp. 2-6 in Performance Standards on Social and Environmental Sustainability. International Finance Corporation, Washington, DC. April 30, 2006 [online]. Available: http://www.ifc.org /ifcext/enviro.nsf/AttachmentsByTitle/pol_PerformanceStandards2006_full/$FILE /IFC+Performance+Standards.pdf [accessed May 23, 2011].

IPCS (International Programme on Chemical Safety). 2006. Draft Guidance Document on Characterizing and Communicating Uncertainty of Exposure Assessment, Draft for Public Review. IPCS Project on the Harmonization of Approaches to the Assessment of Risk from Exposure to Chemicals. Geneva: World Health Organization [online]. Available: http://www.who.int/ipcs/methods/harmonization/areas/dr aftundertainty.pdf [accessed Aug. 10, 2011].

Kakwani, N., A. Wagstaff, and E. van Doorslaer. 1997. Socio-economic inequalities in health: Measurement computation, and statistical inference. J. Econometrics 77:87-103.

Kearney, M. 2004. Walking the walk? Community participation in HIA: A qualitative interview study. Environ. Impact Assess. Rev. 24(2):217-229.

Kwiatkowski, R.E. 2011. Indigenous community based participatory research and health impact assessment: A Canadian example. Environ. Impact Assess. Rev. 31(4):445-450.

Kwiatkowski, R.E., C. Tikhonov, D.M. Peace, and C. Bourassa. 2009. Canadian indigenous engagement and capacity building in health impact assessment. IAPA 27(1):57-67.

Lexchin, J., L.A. Bero, B. Djulbegovic, and O. Clark. 2003. Pharmaceutical industry sponsorship and research outcome and quality: Systematic review. BMJ 326(7400):1167-1170.

Luber, G., and N. Prudent. 2009. Climate change and human health. Trans. Am. Clin. Climatol. Assoc. 120:113-117.

Mahoney, M.E., J.L. Potter, and R.S. Marsh. 2007. Community participation in HIA: Discords in teleology and terminology. Crit. Public Health 17(3):229-241.

Milner, S.J., C. Bailey, and J. Deans. 2003. Fit for purpose' health impact assessment: A realistic way forward. Public Health 117(5):295-300.

Mindell, J., A. Hansell, D. Morrison, M. Douglas, and M. Joffe. 2001. What do we need for robust, quantitative health impact assessment? J. Public Health Med. 23(3):173-178.

Mindell, J., A. Boaz, M. Joffe, S. Curtis, and M. Birley. 2004. Enhancing the evidence base for health impact assessment. J. Epidemiol. Community Health 58(7):546-551.

Mindell, J., J. Biddulph, L. Taylor, K. Lock, A. Boaz, M. Joffe, and S. Curtis. 2010. Improving the use of evidence in health impact assessment. Bull. World Health Organ. 88(7):543-550.

Mitchell, G. 2005. Forecasting environmental equity: Air quality responses to road user charging in Leeds, UK. J. Environ. Manage. 77(3):212-226.

Morgan, R.K. 1998. Environmental Impact Assessment: A Methodological Perspective. Boston: Kluwer.

Morgan, R.K. 2011. Health and impact assessment: Are we seeing closer integration? Environ. Impact Assess. Rev. 31(4):404-411.

Murphy, G. 2001. The Constitution of the Iroquois Nations: The Great Binding Law, Gayanashagowa. Iroquois Confederacy and the US Constitution Website. National

Public Telecomputing Network [online]. Available: http://www.iroquoisd/emocracy.pdx.edu/html/greatlaw.html [accessed Nov. 27, 2010].
NIH (National Institutes of Health). 2011. NIH Consensus Development Program. U.S. Department of Health and Human Services, National Institutes of Health [online]. Available: http://consensus.nih.gov/ [accessed Feb. 7, 2011].
NRC (National Research Council). 1983. Risk Assessment in the Federal Government: Managing the Process. Washington, DC: National Academy Press.
NRC (National Research Council). 1989. Improving Risk Communication. Washington, DC: National Academy Press.
NRC (National Research Council). 1996. Understanding Risk: Informing Decisions in a Democratic Society. Washington, D.C.: National Academies Press.
NRC (National Research Council). 2007. Models in Environmental Regulatory Decision Making. Washington, DC: The National Academies Press.
NRC (National Research Council). 2009. Science and Decisions: Advancing Risk Assessment. Washington, DC: National Academies Press.
O'Connell, E., and F. Hurley. 2009. A review of the strengths and weaknesses of quantitative methods used in health impact assessment. Public Health 123(4):306-310.
Ozonoff, D. 1994. Conceptions and misconceptions about human health impact analysis. Environ. Impact Assess. Rev. 14(5/6):499-515.
Parry, J.M., and J.R. Kemm. 2005. Criteria for use in the evaluation of health impact assessments. Public Health 119(12):1122-1129.
Pate-Cornell, M.E. 1996. Uncertainties in risk analysis: Six levels of treatment. Reliab. Eng. Syst. Safe. 54(2):95-111.
PCCRARM (Presidential/Congressional Commission on Risk Assessment and Risk Management). 1997. Framework for Environmental Health Risk Management – Final Report, Vol. 1. [online]. Available: http://www.riskworld.com/nreports/1997/risk-rpt/pdf/EPAJAN.PDF [accessed July 28, 2011].
Petticrew, M. 2007. "More research needed": Plugging gaps in the evidence base on health inequalities. Eur. J. Public Health 17(5):411-413.
Quigley, R., L. den Broeder, P. Furu, A. Bond, B. Cave, and R. Bos. 2006. Health Impact Assessment: International Best Practice Principles. Special Publication Series No. 5. Fargo: International Association for Impact Assessment. September 2006 [online]. Available: http://www.iaia.org/publicdocuments/special-publications/SP5.pdf [accessed May 6, 2011].
Rothman, K.J., and S. Greenland. 1998. Causality and causal inference. Pp. 7-28 in Modern Epidemiology. Philadelphia, PA: Lippincott-Raven.
RWJF/Pew (The Robert Wood Johnson Foundation and The Pew Charitable Trusts). 2011. Health Impact Project [online]. Available: http://www.healthimpactproject.org/ [accessed May 25, 2011].
Schram-Bijkerk, D., E. van Kempen, A.B. Knol, H. Kruize, B. Staatsen, and I. van Kamp. 2009. Quantitative health impact assessment of transport policies: Two simulations related to speed limit reduction and traffic re-allocation in the Netherlands. Occup. Environ. Med. 66(10):691-698.
Steinemann, A. 2000. Rethinking human health impact assessment. Environ. Impact Assess. Rev. 20(6):627-645.
Susser, M. 1986. Rules of inference in epidemiology. Regul. Toxicol. Pharmacol. 6(2):116-128.
UNECE/REC (United Nations Economic Commission for Europe and Regional Environmental Center for Central and Eastern Europe). 2007. Annex A5.2. Description of selected public participation tools. Pp. 197-206 in Protocol on SEA: Resource

Manual to Support Application of the UNECE Protocol on Strategic Environmental Assessment. Draft Final, April 2007. U. N. Economic Commission for Europe, and Regional Environmental Center for Central and Eastern Europe [online]. Available: http://www.unece.org/env/eia/sea_manual/documents/SEAmanualDraftFinalApril2007.pdf [accessed May 23, 2011].

Veerman, J.L., J.J. Barendregt, and J.P. Mackenbach. 2005. Quantitative health impact assessment: Current practice and future directions. J. Epidemiol. Community Health 59(5):361-370.

Veerman, J.L., M.P. Bekker, and J.P. Mackenbach. 2006. Health impact assessment and advocacy: A challenging combination. Soz. Praventivmed. 51(3):151-152.

Veerman, J.L., J.P. Mackenbach, and J.J. Barendregt. 2007. Validity of predictions in health impact assessment. J. Epidemiol. Community Health 61(4):362-366.

Weed, D.L. 2005. Weight of evidence: A review of concept and methods. Risk Anal. 25(6):545-1557.

Wernham, A. 2007. Inupiat health and proposed Alaskan oil development: Results of the first Integrated Health Impact Assessment/Environmental Impact Statement of proposed oil development on Alaska's North Slope. EcoHealth 4(4):500-513.

Wernham, A. 2009. Building a statewide health impact assessment program: A case study from Alaska. Northwest Public Health 26(1):16-17.

WHO (World Health Organization). 1946. Preamble to the Constitution of the World Health Organization as adopted by the International Health Conference, 19-22 June, 1946, New York; signed on 22 July 1946 by the representatives of 61 States. Official Records of the WHO 2:100.

WHO (World Health Organization). 1999. Health Impact Assessment: Main Concepts and Suggested Approaches-the Gothenburg Consensus Paper. Brussels: European Centre for Health Policy, WHO Regional Office for Europe.

Wier, M., J. Weintraub, E.H. Humphreys, E. Seto, and R. Bhatia. 2009. An area-level model of vehicle-pedestrian injury collisions with implications for land use and transportation planning. Accid. Anal. Prev. 41(1):137-145.

Williams, T. 2010. Presentation at the Third Meeting on Health Impact Assessment, July 28, 2010, Irvine, CA.

Wismar, M., J. Blau, K. Ernst, and J. Figueras. 2007. The Effectiveness of Health Impact Assessment: Scope and Limitations of Supporting Decision-Making in Europe. Copenhagen: World Health Organization [online]. Available: http://www.euro.who.int/__data/assets/pdf_file/0003/98283/E90794.pdf [accessed May 18, 2011].

WWF/ZSL/Global Footprint Network (World Wildlife Fund, Zoological Society of London, and Global Footprint Network). 2010. Living Planet Report 2010: Biodiversity, Biocapacity and Development. Gland, Switzerland: WWF [online]. Available: http://wwf.panda.org/about_our_earth/all_publications/living_planet_report/ [accessed May 23, 2011].

Yost, N.C. 2003. NEPA Deskbook, 3rd Ed. Washington, DC: Environmental Law Institute.

5

Structures and Policies for Promoting Health Impact Assessment

The nation's highest priorities for health, as articulated in the Healthy People 2020 initiative, include increasing quality and longevity of a life and eliminating health disparities between sexes, classes, races, and ethnic groups (DHHS 2010). Poor health severely undermines a person's quality of life and places substantial economic burdens on individuals and on society at large. Chapter 2 documents the direct and indirect associations between current health problems and social, economic, and environmental conditions in the United States. It also illustrates how decisions about policies, programs, projects, and plans—especially those emanating from nonhealth sectors—contribute to conditions that influence the public's health. Thus, improving public health substantially will require focused efforts to recognize and address the health implications of decisions made at all levels and in all sectors of government—that is, to incorporate health into policy-making, planning, and decision-making.

Health impact assessment (HIA), an emerging practice in the United States, is one approach for promoting health and disease-prevention objectives. As described in Chapter 3, HIA aspires to assist policy-makers, decision-makers, and the public in identifying health considerations and factoring them into proposed policies, plans, programs, and projects that otherwise would not have recognized or addressed important health risks or opportunities. It aims to protect and promote public health and to reduce health disparities by informing decision-making, and it offers substantial potential benefits to improve public health. In contrast with its more extensive use internationally, HIA appears to be underused in the United States. The committee identified several barriers to the development and use of HIA in the United States:

- The context within which HIA is practiced poses a challenge. There are few legal mandates for the use of HIA in the United States; as described in Chapter 4, the decision-making contexts within which HIA must occur are diverse; and the minimal attention to health in public policy-making has not been identified as a pressing issue on local, state, or national policy agendas.

- Societal awareness of the many determinants of health is limited. The general public and people in a variety of nonhealth (and health) sectors often have little understanding of the influence of all the social, cultural, political, economic, and environmental determinants on health and therefore have little awareness about the utility of HIA. As a result, there is little public demand for the use of HIA in the United States.
- Another key challenge is related to the professional practice of HIA itself. Little education and training in HIA are available in the United States. The current practice of HIA is inconsistent and nonstandardized. The quality of analytic methods used by HIA practitioners varies widely and there is not enough synthesized evidence on health determinants that can be used by HIA practitioners. In addition, the effectiveness of HIA and its effects on public-health outcomes have not been evaluated sufficiently.
- Finally, there are few resources to support the practice of HIA.

In response to those barriers, the committee identified four core issues that must be addressed to foster the judicious, deliberative, and rigorous use of HIA in the United States:

- Structure and policies to support HIA.
- Promotion of education, training, and societal awareness of HIA.
- Increase in research and scholarship in HIA.
- Development of resources to support HIA.

STRUCTURE AND POLICIES TO SUPPORT HEALTH IMPACT ASSESSMENT

The continuing adoption and effectiveness of HIA in the United States are predicated on the creation of an institutional framework that facilitates its use in public decision-making at all levels of government (see Appendix A for international examples of the use of HIA at various levels of government). Although there are a number of ways for such a framework to emerge, two potential ways to support HIA are greater and sustained interagency collaboration among government agencies at local, state, and federal levels and better implementation of existing policies with the creation or strengthening of enabling legislation at local, state, and federal levels.

Interagency Collaboration

It is difficult or impossible to conduct an HIA of policies, programs, and projects of nonhealth public sectors—such as economic policies, job-training programs, and infrastructure projects—without substantial interagency collaboration among sectors and all levels of government. For example, if an HIA of a proposed road expansion is led by a public-health agency, the HIA team will need to work with the departments of public works, planning, and engineering to understand the proposed project fully. Conversely, if the HIA is led by a non-health agency, the HIA team will need input from a public-health agency on relevant health data. In short, the practice of HIA depends on and benefits from cross-agency collaboration. Such collaboration is also essential because of the resource-constrained environment within which public-policy-makers and public officials work.

Although the nature and extent of collaboration will depend on the level of government and the particular decision context, the collaborative arrangements—which may be manifested in joint task forces, councils, cabinets, new departments, shared staff appointments, or some other suitable mechanism—are most effective when they represent the widest possible group of professional interests, such as departments of public health, planning, law, and economic development.

There are a number of potential ways to promote interagency collaboration. The committee notes several examples below.

- Federal agencies, such as the Council for Environmental Quality (CEQ) and the U.S. Centers for Disease Control and Prevention (CDC), could establish collaborative relationships—for example, through an interagency working group or a task force—that would be explicitly charged with developing guidance for integrating health concerns into the implementation of the National Environmental Policy Act (NEPA). Existing regulations that provide a foundation for such guidance are discussed in Appendix F.
- Individual executive-branch agencies could evaluate whether HIA is an appropriate mechanism for incorporating health considerations into their plans and proposals and for meeting standards conferred by their enabling legislation and regulations concerning public health and well-being.
- The Affordable Health Care for America Act of 2009 set out objectives for the member agencies of the National Prevention, Health Promotion, and Public Health Council (2010). The council could consider how HIA might be used to achieve those objectives, and it could also recommend use of HIA in the National Prevention and Health Promotion Strategy.
- Tribal health departments could become involved in NEPA-related decisions made by federal agencies when it appears that decisions would be important for tribal health or well-being. There are several opportunities for tribal participation in the NEPA process. First, tribal members and government representatives can submit formal comments. Second, tribal governments may request direct "government-to-government" consultation with lead federal officials at any time during the NEPA process. Third, tribal governments may ask to become "cooperating agencies" in the preparation of NEPA-related documents;

this role allows them to review, comment on, and contribute new information to the analysis as it is being developed.

- Tribes could consider forming multiagency working groups to locate appropriate opportunities to incorporate health into planning, policy, and programmatic decision-making.
- As in efforts at the federal level, state health departments and departments of the environment could establish interagency working groups charged with integrating health concerns into decision-making processes at the state level.
- State agencies—such as departments of the environment, agriculture, education, and transportation—could invite their health departments to participate in coordinated planning and permitting activities for large projects and for infrastructure or transportation improvement programs. This approach is proving successful in at least one state (Wernham 2009; Health Impact Project 2010).
- Local public-health agencies—county and city health departments—could partner with other government agencies, such as agencies of urban planning and economic development, in promoting health. HIA could be used as a tool to engage the agencies. This practice has shown considerable promise in several jurisdictions (Bhatia and Wernham 2008; Corburn 2009).
- Local public-health agencies could become more multidisciplinary by deepening expertise in nonhealth sectors and could assist in building capacity in other agencies. For example, public-health agencies might train planners and other officials in the use of HIA.
- Given the sparse resources of local government agencies, innovative revenue-generation options will need to be explored to support many of the above activities. For example, health departments that are involved in formal planning or permitting decisions could be funded by such mechanisms as permitting fees.

Supportive Public Policies and Legislation

HIA can be advanced by fully implementing existing policies and legislation that support the use of HIA or through support of the creation of new enabling legislation. The key policies that support the use of HIA in the United States are NEPA and state environmental policy acts (SEPAs) (see Appendix A for further discussion). Although the federal NEPA process and equivalent processes at the state level are important tools for advancing HIA, it is possible and probably prudent for the public sector to enact additional policies and legislation outside the context of NEPA and SEPAs to facilitate the use of HIA. Making prescriptive recommendations on the nature of the new policies and legislation is beyond the scope of this report, particularly given the wide variation in policy contexts across the country. Instead, several avenues through which HIA may be advanced are outlined below; some of which focus on reinvigorating and strengthening the spirit of NEPA. The examples are by no means exhaustive; they constitute only a sample of general approaches that could be used to further the practice of HIA.

- Explicit guidance that demonstrates how health considerations can be incorporated into NEPA could be developed jointly by agencies best suited to the task of integrating health into the NEPA process and provided to federal agencies. For example, CEQ in partnership with CDC and other appropriate public-health and NEPA experts could develop and issue guidance to federal agencies on explicitly incorporating health considerations into NEPA. The guidance could also encourage lead federal agencies to solicit appropriate participation of local, state, tribal, or federal health officials as cooperating agencies in the NEPA process.
- Without clear health goals, objectives, metrics or indicators, or targets, it is difficult for federal agencies to gauge and monitor the extent to which health and HIA are incorporated into policies. One possibility is for federal agencies to develop such metrics and targets as part of their 5- and 10-year plans. The metrics could be adopted from the Healthy People 2020 initiative, which provides science-based 10-year objectives for measuring improvements in health (DHHS 2010). Such an approach is consistent with the framework of Healthy People 2020, which argues for a "health in all policies" approach.
- To overcome institutional barriers, it is important to identify means to facilitate the explicit inclusion of health concerns in domestic policy-making at all levels of government. One strategy for doing so could be the establishment of a committee, council, or task force nested within existing policy-making bodies at the federal level (such as the Domestic Policy Council) with analogues at the state and local levels. To be successful, such an entity would need to have clear points of coordination at all levels of government, identifiable liaisons, and a clearly defined charge.
- The Government Accountability Office could review, synthesize, evaluate, and publically disseminate information on HIAs of federal government policies, projects, and programs.
- The U.S. Environmental Protection Agency could consider ways of expanding their reviews of environmental impact statements to include assessment of health consequences for low-income populations, racial minorities, and native tribes (42 U.S.C. Section 7609 (1970)).
- Each policy sector—such as energy, housing, and transportation—could consider including explicit objectives and performance measures in planning, funding, and policy-development activities that are aimed at protecting human health. For example, the transportation sector could include planning and design objectives that would result in reduction of human exposure to air pollution and prevention of injuries to pedestrians, bicyclists, and other users of roads. The housing sector could include objectives and measures for reducing segregation, crowding, and injury hazards.

- Tribal governments could consider enacting a tribal environmental policy act and include standards for the use of HIA when appropriate (Tulalip Tribes 2000).

At the local government level, HIA may be useful as a tool for reviewing the effects of plans and projects on the health of a community. Several examples are noted below.

- HIA may be used to gauge the effect of comprehensive plans on the health of a community, especially in cases in which health is not explicitly an element of a local comprehensive plan.
- One purpose of zoning is to protect public health and well-being. HIA is proving to be a useful tool for assessing the effects of proposed new or revised zoning codes on public health.
- School districts could use HIAs to gauge the effects of various discipline policies, exercise curricula, school-meal programs, or school-siting decisions on children's health. Health and wellness committees in school districts can play a key role in initiating a conversation around HIAs.

PROMOTION OF EDUCATION AND TRAINING IN AND SOCIETAL AWARENESS OF HEALTH IMPACT ASSESSMENT

A few institutions of higher learning in the United States offer formal education in HIA; for example, the University of Wisconsin-Madison and the University of California, Berkeley offer courses that feature HIA. Other courses are taught by practitioners in the field. For example, the San Francisco Department of Public Health has taught an annual 4-day course for practitioners for the last few years, and several other organizations—such as Human Impact Partners, Design for Health, CDC, and the University of California, Los Angeles—have offered training (usually for 1-3 days) and technical assistance.

Few professionals in the United States, however, are trained in the practice of HIA. Current HIA practice in the United States is based largely on experiential learning, that is, "learning by doing." The present committee views high-quality education and training as critical for the advancement of HIA in the United States. The committee notes that advancement must occur in basic education, continuing education, and formation of professional associations.

Basic Education in Health Impact Assessment

HIA is concerned with bringing health concerns into a decision-making process that would otherwise fail to incorporate health. Therefore, HIA practitioners will always work in interdisciplinary settings and with interdisciplinary groups, and the education of future HIA professionals in academic settings must embody a variety of relevant disciplines—health-related (such as public health and medicine) and other (such as public policy, urban planning, public administration, and economics). The teaching must engage faculty and students in the various disciplines. Accordingly, schools of public health and medicine, public policy, urban planning, public administration, and economics should develop curricula that enable studies to learn core HIA skills. The curriculum must adhere to the highest standards of academic rigor as demanded by the core disciplines in which HIA is taught.

Material, financial, human, and institutional resources are necessary from inside and outside academe to facilitate inclusion of HIA in academic programs. Potential agencies outside the academic setting that might support educational programs in HIA are those whose mission is to promote health (such as the National Institute of Environmental Health Sciences) and education in general (such as the U.S. Department of Education).

Continuing Education of Professionals, Policy-Makers, and Society

In addition to introducing HIA into academic programs, the committee views continuing education of HIA professionals, policy-makers, and society in general as important for improving the quality of HIA practice in the country. It is especially important to emphasize broad societal education in the many determinants of health so that individuals and communities can make informed decisions about their health and well-being and can participate fully in the HIA process.

One possibility for promoting continuing education of professionals is flexible and modular training programs in a variety of agencies—public, nonprofit, and private—and in different levels of government. For example, the CDC Healthy Community Design Initiative has supported state health departments in training and mentoring local health departments in HIA; the initiative has made it possible for several jurisdictions to complete HIAs (CDC 2011). Such training should be expanded to reach a wider array of individuals and groups. Furthermore, because HIA practice has to overcome barriers related to the lack of interagency collaborative structures, it is important to engage and train senior-level local, state, and federal agency officials and decision-makers. Leaders of the federal civilian workforce, such as the federal Senior Executive Service (OPM 2011), could benefit from continued education in HIA because it would raise health awareness in their own work.

Emergence of Professional Associations and Groups

Like any growing field, the field of HIA could benefit from a professional association or society. The society could facilitate continued professional development of HIA practitioners and develop, monitor, and facilitate standards of professional education and practice in HIA. It could also establish and oversee publication of peer-reviewed research and scholarship in and about HIA through a professional journal. Since 2008, a network of practitioners in North America has been working to advance the practice of HIA in the United States. The first collective product of the network was a set of minimum elements and voluntary practice standards for the field (Bhatia et al. 2010). The network has continued to meet periodically and is taking steps to build awareness, mentor new practitioners, and support integration of HIA and EIA. It is expected to formalize its relationships and activities in a professional organization in the near future.

INCREASE IN RESEARCH AND SCHOLARSHIP IN HEALTH IMPACT ASSESSMENT

Scholarship for Developing Methods and Evidence for Health Impact Assessment

The methods and evidence used in HIA practice vary widely and are inconsistent in quality. Research to improve the analytic methods available to HIA practitioners is important, and research evidence that ties distal upstream factors to health outcomes that could be used in the HIA process is essential. Suggested research topics on the role of distal or upstream factors[1] in health that could strengthen the evidence base available to HIA practitioners include the following:

- How health is affected by specific federal policy decisions and actions related to agricultural policy, education, energy development, environmental protection, housing, immigration, infrastructure, military defense, national parks, natural resources, taxation, and transportation.
- How health is affected by state fiscal policy (such as property tax law), agriculture, education, welfare-to-work, and land-use and growth-management policies.[2]
- How health is affected by planning processes (such as comprehensive planning, growth-management planning, and land-use planning), regulatory mechanisms (such as subdivision regulations and zoning and building bylaws), fiscal tools (such as local tax regulations and incentives), infrastructure projects, and school district policies.

Beyond the primary research suggested above, HIA practice would also be enhanced by developing approaches to apply decision-theory concepts in the context of the complex quantitative and qualitative information used in HIA. Evaluating multiple alternative policies in the face of tradeoffs and uncertainty is the

---

[1]These factors include the role of the natural and built environments and social, economic, and political environments in fostering or hindering public health.

[2]Not all states in the country enact statewide land-use and growth-management policies. In states where such policies exist, consideration of HIA is relevant.

hallmark of decision science, and methods that can leverage the strengths of decision-science approaches—such as multiattribute utility analysis (Keeney and Raiffa 1976)—in the context of HIA would be valuable (Merkhofer et al. 1997).

Scholarship on Health Impact Assessment Practices and Their Effectiveness

Evaluation of HIA has occurred to some extent internationally (Harris-Roxas 2009). However, because HIA is relatively new in the United States, there is a paucity of evidence on the effectiveness of HIA practice in this country. Such research is especially necessary inasmuch as HIA may require the investment of substantial public and private resources. Research is needed to document HIA practices and its effectiveness in influencing decision-making processes and promoting public health. Existing tools of evaluation research might be used and adapted to evaluate HIA (Rossi et al. 1999; Trochim 2000). Potential research includes the following:

- Development and empirical validation of theories or frameworks to understand and assess the effect of HIA on decision-making and related social processes.
- The effect of HIA on improving short-term and long-term health outcomes.
- The role of local, tribal, state, and federal governance structures and decision-making processes in integrating public-health concerns into public policy.
- Methods to address the challenges and opportunities in using HIA to inform government decision-making at all levels and branches of government.

Improvement of HIA practice requires scholarship for and on HIA practice, and such scholarship cannot be generated without financial support. Financial support can come from philanthropic, private, and public entities, such as the National Institutes of Health, CDC, and the Agency for Healthcare Research and Quality.

HIA practitioners are most likely to benefit from translational research that synthesizes high-quality scientific evidence for use by practitioners and policymakers. Such an effort would have to gauge the quality of the latest available research evidence on the role of distal factors on public health and synthesize that information for use by HIA practitioners.

The synthesized evidence can be disseminated to practitioners by using a variety of tools, such as journals, on-line repositories, and newsletters. Among those options, an online repository would be a centralized and dynamic tool for bringing the latest synthesized research to HIA practitioners. Such a repository may be made available by a number of entities, including universities, research centers, private groups, and government agencies, such as CDC.[3] As a publicly available and credible source of information on public health for the nation, CDC is especially well-positioned to establish and maintain such a repository.

DEVELOPMENT OF RESOURCES TO SUPPORT HEALTH IMPACT ASSESSMENT

A key barrier to the use of HIA is the availability of resources for communities and groups interested in undertaking it. Resources are also essential for continued education and training of professionals in the field, and the lack of resources affects the quality of HIA. Furthermore, resources are needed for monitoring and conducting evaluations.

For more resources to become available to support the development of HIA practice, society as a whole has to recognize the importance of considering health in all policies, programs, plans, and projects to improve quality of life and to protect the health of future generations. Yet, many of the connections that HIA makes explicit are neither obvious nor intuitive to the general public or to decision-makers in nonhealth (and health) agencies. A national information campaign is crucial for highlighting the importance of a wide array of decisions to public health, clarifying the role of HIA in the decision-making process, and advancing HIA practice. Such a campaign could be conducted by existing health agencies, such as CDC, or by new organizations, such as a new association for HIA, if such an entity were to emerge. Such information could be disseminated through an online repository, for example, one managed by CDC.

Although this chapter is focused largely on barriers to and options for developing structures and policies to support HIA in the public sector, the committee recognizes that private-sector decisions also have health implications. The committee encourages the private sector to incorporate HIA into projects and developments that are likely to have important impacts on health and health determinants. Private-sector planning and development initiatives could also consider using HIA as a means of informing stakeholders of possible adverse or beneficial effects and allowing them to participate in planning and shaping proposed projects, programs, or plans in a way so as to minimize adverse effects and optimize beneficial ones.

REFERENCES

Bhatia, R., and A. Wernham. 2008. Integrating human health into environmental impact

---

[3] A number of on-line resources for HIA exist; for example, the University of California, Los Angeles offers an on-line learning and information center on HIA, and the Health Impact Project offers an interactive, searchable database of completed and in-progress HIAs in the United States. However, providing a synthesis of research evidence does not appear to be the central function of such Web sites.

assessment: An unrealized opportunity for environmental health and justice. Environ. Health Perspect. 116(8): 991-1000.

Bhatia, R., J. Branscomb, L. Farhang, M. Lee, M. Orenstein, and M. Richardson. 2010. Minimum Elements and Practice Standards for Health Impact Assessment (HIA), Version 2. North American HIA Practice Standards Working Group, Oakland, CA. November 2010 [online]. Available: http://www.sfphes.org/HIA_Tools/HIA_Practice_Standards.pdf [accessed May 23, 2011].

CDC (Centers for Disease Control and Prevention). 2011. Healthy Community Design Initiative: Recent Accomplishments. Centers for Disease Control and Prevention [online]. Available: http://www.cdc.gov/healthyplaces/accomplishments.htm [accessed July 25, 2011].

Corburn, J. 2009. Toward the Healthy City: People, Places and the Politics of Urban Planning. Cambridge: The MIT Press.

DHHS (U.S. Department of Health and Human Services). 2010. Health People 2020 Framework. Office of Disease Prevention and Health Promotion, U.S. Department of Health and Human Services [online]. Available: http://www.healthypeople.gov/2020/consortium/HP2020Framework.pdf [accessed Feb. 2, 2011].

Harris-Roxas, B. 2009. Conceptual Framework for Evaluating the Impact and Effectiveness of Health Impact Assessment. Centre for Health Equity Training, Research and Evaluation (CHETRE), The University of New South Wales, Sydney [online]. Available: http://www.hiaconnect.edu.au/evaluating_hia.htm [accessed May 25, 2011].

Health Impact Project. 2010. Alaska Department of Health and Social Services Seeks to Hire a Medical Epidemiologist for HIA Program. Health Impact Project In the News: January 12, 2010 [online]. Available: http://www.healthimpactproject.org/news/in/alaska-department-of-health-and-social-services-seeks-to-hire-a-medical-epidemiologist-for-hia-program [accessed July 25, 2011].

Keeney, R.L. and H. Raiffa. 1976. Decisions with Multiple Objectives: Preferences and Value Tradeoffs. Hoboken, NJ: John Wiley and Sons.

Merkhofer, M.W., R. Conway, and R.G. Anderson. 1997. Multiattribute utility analysis as a framework for public participation in siting a hazardous waste management facility. Environ. Manage. 21(6):831-839.

National Prevention, Health Promotion and Public Health Council. 2010. Status Report. July 1, 2010 [online]. Available: http://www.hhs.gov/news/reports/nationaprevention2010report.pdf [accessed Feb. 2, 2011].

OPM (U.S. Office of Personnel Management). 2011. About the Senior Executive Service. U.S. Office of Personnel Management, Washington, DC [online]. Available: http://www.opm.gov/ses/about_ses/index.asp [accessed Feb. 3, 2011].

Rossi, P.H., H. Freeman, and M.W. Lipsey. 1999. Evaluation, Sixth edition. Thousand Oaks, CA: Sage Publications.

Trochim, W. 2000. The Research Methods Knowledge Base, 2nd Edition. Cincinnati, OH: Atomic Dog Publishing.

Tulalip Tribes. 2000. A Comprehensive Guide for American Indian and Alaska Native Communities. The Tulalip Tribes of Washington Present: Participating in the National Environmental Policy Act and Developing a Tribal Environmental Policy Act. October 2000 [online]. Available: http://knowledge.fhwa.dot.gov/ReNEPA/ReNepa.nsf/All+Documents/C3A140A5BC48BC8D852570240073CFA3/$FILE/TEPA.pdf [accessed July 25, 2011].

Wernham, A. 2009. Building a statewide health impact assessment program: A case study from Alaska. Northwest Public Health 26(1):16-17.

Appendix A

Experiences with Health Impact Assessment

To develop a framework and guidance for the practice of health impact assessment (HIA) in the United States, the committee felt that it was critical to review the HIA experience of the international community given its use of HIA over the last several decades. The international experience in implementing HIA has involved different institutional arrangements, mechanisms for knowledge transfer, tools, and capacity. On examination of the international experience, the committee identified three main mechanisms for introducing HIA. The first is to incorporate HIA into existing assessment processes—for example, environmental impact assessment (EIA) under the National Environmental Policy Act (NEPA)—and thus make human health an explicit consideration in the mechanisms for approval of policies, plans, programs, and projects. The second is to require HIA explicitly by law or regulation or in response to defined triggers. The third is to use HIA voluntarily but to provide various degrees of government support and resources. In this appendix, the committee examines how the international community has used those mechanisms and what lessons the global experience offers for one who is considering a framework and guidance for HIA in the United States.

This appendix is not a comprehensive review, but it seeks to summarize HIA experience in Canada, Europe, Australia, and Thailand. It also looks at the use of HIA by indigenous people and multilateral organizations. The committee reviews HIA experience in the United States and discusses the relationship between HIA and the process of EIA. The appendix concludes with comments on the use of HIA in the private sector and some important lessons learned from the experience to date that are relevant to the future use of HIA in the United States. The committee notes that this appendix uses the terms *health* and *health impact assessment*. To examine the international experience, the committee recognized that it was important to consider the wider policy context and to view HIA as one among many methods by which health is incorporated into decision-making.

CANADA

In the early 1970s, a central government think tank, the Long Range Health Planning Branch, identified the effects of lifestyle and environment on public health and began to consider policy solutions to improve public health (Laframboise 1973; McKay 2000). That activity culminated in a report that identified objectives for the health-care system and for the prevention of health problems and promotion of good health (Lalonde 1974). A combination of research and advocacy was introduced to support and validate the notion that public policies affect determinants of health (Milio 1981; WHO 1986, 1988).

Healthy Public Policy

Health and environment are under provincial jurisdiction in Canada. Two provinces, British Columbia and Québec, have formalized HIA as a component of policy-making, and they offer different experiences (Banken 2001, 2004; Kwiatkowski 2004; Gagnon et al. 2008). In British Columbia, attention to the health of the population was advanced by a group of government officials who had an interest in health promotion. From 1989 to 1995, structures and policies for HIA were starting to be included in British Columbia's health-care policy, and it was proposed that HIA of all government projects, programs, and laws be conducted. Guidelines were produced, and a series of workshops were held to raise awareness of and develop capacity for HIA[1] (Banken 2004). By 1999, the values underpinning the reform of health care had changed, and resources for HIA were redeployed. The guidelines that required the use of HIA in government decisions were not changed, but they were no longer seen as mandatory. Banken (2004) concluded that the rise of HIA in that short time had been accelerated by key persons in the British Columbia Ministry of Health, that it did not benefit from wide ownership, and that it had become closely identified with a particular policy orientation. Banken contended that if other institutions had been more involved in examining the value of and establishing structures for HIA, support for HIA would not have withered as quickly after the policy direction changed and after key persons left the ministry.

Québec had a different experience in using HIA as part of healthy public policy. Banken (2001) traced the linking of environment and health to robust public-health input during hearings on the use of pesticides (BAPE 1983). That input led to a memorandum of understanding (MOU) between Québec's Ministries of Health and Environment. A framework was developed to support the memorandum and led to the systematic practice of integrating health and the environment into projects and policies (Banken 2001, 2004). In the 1990s, pol-

---

[1]The committee is not aware of any examples of HIA from this period. Therefore, although it is documented that HIA was a part of the policy discussion, it is not possible to evaluate how HIA was conducted in British Columbia.

icy documents recognized the need for intersectoral initiatives to improve health (Government of Québec 1998, 1999) and explicitly recommended the systematic assessment of the impacts of public policies on health. The assessments were to be conducted by the Study Commission for Health and Social Services (Commission d'Étude sur les Services de Santé et les Services Sociaux), which analyzes health services.

HIA was included in Québec's 2001 Public Health Act, which requires government ministries and agencies to ensure that legislative provisions do not adversely affect the health of the population. It also requires that the minister of public health be consulted on all policies that could have an important health effect (Section 54, Government of Québec 2001). Figure A-1 shows the number of requests for consultations from other ministries. In 2011, the national public health director and the assistant deputy minister in the Ministry of Health and Social Services (Ministère de la Santé et des Services Sociaux) of Québec stated that there were 434 requests for advice from 2003 to 2011 (Poirier 2011a). Although the demands of the legislative calendar influence the number of requests from year to year, the figure indicates a clear upward trend. The trend is ascribed to the Ministry of Health and Social Service's efforts to develop an understanding of Section 54 across the government, improvements in how the ministry processes requests for consultation and provides its advice, and the application of a public-health perspective to a wider array of policies.

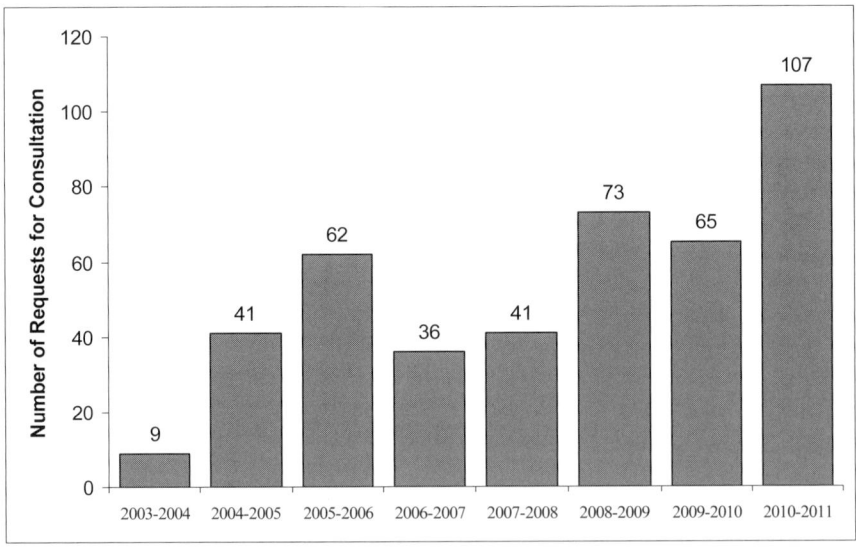

FIGURE A-1 Number of requests for consultation received by the Québec Ministry of Health and Social Services, 2003-2008. Source: L. Jobin, Ministry of Health and Social Services, Québec, personal communication, 2011.

The Québec public-health law is noteworthy because it focuses on the processes by which the government will request assistance on health issues and on how that assistance will be provided by the Ministry of Health (for more information, see NCCHPP 2008). Clearly defining the process has helped to ensure that government departments request health input when writing policy. The Ministry of Health and Social Services has also worked to heighten awareness and to gain the support of other government ministries and agencies (NCCHPP 2008). Although changes are occurring slowly, channels of communication beyond government departments covered by the law are being opened, and this has led to the integration of the health sector (and health consideration) into the political-administrative process. Furthermore, the government's knowledge development and transfer strategy to support the implementation of the law has strengthened research capacity on healthy public policy in academic sectors and in the Institut National de Santé Publique du Québec (L. St-Pierre, National Collaborating Centre for Public Policy and Health, Québec, personal communication, 2010).

Some issues, however, still need to be resolved. Many government ministries do not comply with the law, and most requests to the Ministry of Health come from the Executive Committee, which is well versed in the importance of health effects. In addition, the process does not specify a particular method of conducting a health assessment (L. St-Pierre, National Collaborating Centre for Public Policy and Health, Québec, personal communication, 2010). Further efforts clearly are required to foster responsibility for health in some parts of government, such as economics and finance. The next steps envisaged include feedback mechanisms to monitor and evaluate how support is offered and taken and how recommendations are implemented. Continued support for changes in practice is needed through high-quality and strategic evaluations that facilitate actions early in the decision-making process, knowledge transfer, and strategic monitoring (Héroux de Sève et al. 2008; Poirier 2011b).

Examining Human Health in Environmental Impact Assessment

In 1995, the Federal-Provincial-Territorial Committee on Environmental and Occupational Health convened a task force in response to reviews that demonstrated that health aspects were inconsistently or only partially addressed in EIA. The task force was asked to develop a definition of HIA that would be acceptable to all jurisdictions, a public-health framework appropriate to HIA, guidance and training material for HIA, and strategies for increasing awareness about HIA, EIA, and the relationship between human health and the environment (Kwiatkowski 2004). The task force concluded that HIA should be promoted within the existing legislated federal or provincial EIA processes; that HIA was not the responsibility of any one government department or agency in that many factors—including environmental, social, economic, and occupational ones—affect public health; and that HIA should use a multidisciplinary approach informed by the many determinants of health rather than a narrow definition of health (Kwiatkowski 2004). A review by Davies and Sadler (1997) was influential in establishing a case for examining human health in environmental assessment in Canada. A major output of the initiatives was the *Canadian Handbook on Health Impact Assessment*, a comprehensive resource that was first published in 1998 and has since been updated (Health Canada 2004a,b,c,d).

About 6,000 projects a year undergo EIA under the Canadian Environmental Assessment Act, so it is no small feat to ensure that potential health effects are considered for each project (Kwiatkowski and Ooi 2003). EIAs are characterized as screening, comprehensive study, or public-panel review. As implied by its name, screening is less intensive than the other types and accounts for over 95% of EIAs conducted (Kwiatkowski and Ooi 2003).

What is the current experience of incorporating HIA into EIA? Social effects are considered in EIA in Canada; this makes it somewhat easier to include a wide array of health determinants in assessments (M. Orenstein and M. Lee, Habitat Health Impact Consulting, personal communication, 2011). Noble and Bronson (2005, 2006) reviewed three mining case studies and conducted a survey of environmental-assessment practitioners, health practitioners, administrators, and special-interest groups in northern Canada. They found that health has typically been considered only in the early stages of the environmental-assessment process and that only physical health effects associated with project-related environmental damage have generally been considered. As a rule, health and social determinants have not been considered or have been considered only in the context of factors—such as employment opportunities and worker health and safety—that the project sponsor directly controls. The authors acknowledged, however, that the scope of attention to health in EIA has more recently been expanded to reflect a wider array of health determinants that includes a group's culture and its traditional land use. They concluded that there is a need to adopt measures to mitigate adverse effects and optimize beneficial effects that the community is sensitive to, to ensure that the measures are effective, and to monitor and evaluate the effects after project approval (Noble and Bronson 2005, 2006). The committee notes that the somewhat bleak assessment by the authors is based on a small sample and may be unduly harsh.

Although systematic collaboration between public health and the environment sector can be improved, research indicates that health is being considered to some extent in EIA. Overall, Canada has some of the most extensive and successful experiences of including HIA in EIA and of analyzing and improving HIA practice. This work is not always labeled as HIA, but health is increasingly a component of an integrated approach to environmental assessment (Orenstein et al. 2010; M. Orenstein and M. Lee, Habitat Health Impact Consulting, personal communication, 2011).

THE EUROPEAN UNION

HIA has been practiced in the European Union (EU) since the 1980s. During the 1990s, there were developments in HIA methodology and practice in Germany, the Netherlands, Sweden, and the United Kingdom. In the late 1990s, the WHO European Centre for Health Policy played a key strategic role in European HIA policy development, and its 1999 Gothenburg consensus conference produced the first universally accepted definition of HIA.

Although requirements and practice have differed, there are examples of health assessment in the environmental-assessment framework,[2] in stand-alone HIAs, and in all types of policies—from local policies to policies covering the EU. Explicit policies for HIA exist, but its practice is often advanced through the actions of committed individuals. Research grants from the EU play an important role in enabling research and in developing techniques and capacity for HIA. The grants have funded multicenter studies that involve universities, the public sector, and occasionally private-sector bodies across the EU (see, for example, Abrahams et al. 2004; Hilding-Rydevik et al. 2005; WHO 2005a,b,c,d; Wismar et al. 2007; Gulis et al. 2008; HEIMTSA consortium 2010; and INTARESE consortium 2010).

In the EU, HIA is recognized as a process that sits within the broader sphere of public-health policy and sustainable development. It is one of the ways in which partnerships are developed between municipalities and health authorities and is increasingly used as a mechanism by which land-use or spatial planning can work in partnership with public health. Although skills and capacity for HIA are not widespread, there are isolated examples of universities' incorporating HIA as part of a curriculum to train planners and public-health professionals. In a study of HIA across Europe, Wismar et al. (2007) showed that HIA has been used in various countries, at various levels, and in various sectors. They noted that participation and equity considerations have played substantial roles in the practice of HIA and concluded that despite the reported variations, HIA can be used prospectively, cover all stages of the policy process, and use different types of approaches.

The following sections provide background on the EU and on the integrated assessment framework used for EU policy. Approaches for integrating health into environmental assessment across Europe are discussed next,[3] and then other approaches that have been put into place across Europe to enable HIA to be conducted are reviewed.

---

[2]Regarding environmental assessment in the EU, human-health measures are included in directives and legislation that regulate the effects of development on the environment.

[3]This summary does not examine legislation for equality and human rights in the EU, which also leads to policy assessment and can incorporate health issues.

Incorporation of Health into Policies, Plans, Programs, and Projects in the European Union

In 2010, the EU had 27 member states and four applicants for membership (see Box A-1). Policies and laws that apply throughout the EU are produced mainly by the joint work of three institutions: the European Commission, the European Parliament, and the Council of the European Union. The European Commission, which proposes new laws and then works with member states to implement them, is divided into departments and services (EC 2011a). Public health falls under the Directorate-General for Health and Consumers, and environmental stewardship falls under the Directorate-General for the Environment. Public health is a relatively new policy topic at the EU level, and member states continue to hold the main responsibility for national health policy.[4] Actions at the EU level complement actions at the national level, for example, by addressing major health threats and issues that have a cross-border or international impact, such as pandemics and bioterrorism; by addressing health threats related to the free movement of goods, services, and people; by promoting healthier lifestyles; and by supporting the work of national authorities. It is recognized that public health is not solely an issue for health policy. For example, in 1997, the Amsterdam Treaty of the EU required that all European Community policies protect health. Thus, the "health in all policies" approach is required for internal and external policies, and support is given for the use of impact assessment and other tools that evaluate health (CEC 2007).

The European Commission assesses initiatives for their potential economic, social, and environmental consequences before it proposes them (EC 2011b). Health is considered in that process as one of several topics in an integrated impact assessment framework. The guidelines for the framework were updated in 2009 to review public health and safety and to enhance the consideration of social impacts, including access to and effects on social protection, health, and educational systems (EC 2009a). Specific attention has been given to distributional effects and effects on poverty and social inclusion in the EU and developing countries (EC 2009b). Reviews show a small increase in the number of mentions of the word *health* in the European Commission's impact assessment reports; thus, although progress is slow, consideration of health in the framework is increasing (Ståhl 2010). However, the framework for impact

---

[4]Before 1992, health was addressed in the context of health and safety in the workplace and as an issue of consumer safety. The 1992 Maastricht Treaty (EC 1992) was the first treaty to feature an article on public health and to explain the added value of Europe-wide approaches to common challenges in health while confirming that health care remains the mandate of national authorities. Later reform treaties (EC 1997, 2007) enhanced the role of the EU in supporting member states in cooperating and sharing good practice, such as in health-technology assessment, and in tackling cross-border health threats and disease prevention.

> **BOX A-1** European Union Members and When They Joined
>
> 1952 – Belgium, France, Germany, Italy, Luxembourg, and Netherlands
> 1973 – Denmark, Ireland, and United Kingdom
> 1981 – Greece
> 1986 – Portugal and Spain
> 1995 – Austria, Finland, and Sweden
> 2004 – Cyprus, Czech Republic, Estonia, Hungary, Latvia, Lithuania, Malta, Poland, Slovakia, and Slovenia
> 2007 – Bulgaria and Romania
> Candidate Countries: Croatia, Turkey, the Former Yugoslav Republic of Macedonia, and Iceland.
>
> Source: EU 2011a,b.

assessment has been criticized for failing to improve the consideration of public health, for example, in focusing on specific health services rather than the wider health of the general public (Ståhl 2010), in placing a low priority on health so that it is not seen as a factor that can differentiate between policy options (Ståhl 2010), in focusing on the effects on the economy or the business environment, and in being open to undue influence from corporate interests (Smith et al. 2010a,b).

Environmental-Assessment Directives

As noted, one of the roles of the Directorate-General for the Environment is to ensure that member states comply with the requirements of the environmental directives. Environmental assessment is a key mechanism for evaluating individual projects identified by the EIA directive (Council of the European Union 1985) or public plans or programs identified by the strategic environmental assessment (SEA) directive (EP/Council 2001). "The common principle of both directives is to ensure that plans, programs, and projects that are likely to have significant effects on the environment are made subject to an environmental assessment prior to their approval or authorization" (EC 2011c). Consultation with the public is a key feature of environmental-assessment procedures.

Member states are free to supplement the assessment processes, and they must incorporate them into their national consent regimes (that is, the framework by which projects are given permission). For that reason, there is some variation in processes between member states. The Directorate-General for the Environment ensures that each member state implements the EIA and SEA directives, and the European Court of Justice is the final arbiter if assessments are disputed. As both directives are procedural, the courts tend to be concerned with how the assessments have been conducted rather than with their accuracy. Issues of quality are typically left to the organizations overseeing the consenting process, although that can be problematic; for example, health authorities are not always asked to comment on the health components of environmental assessments.

Environmental Impact Assessment

The EIA directive applies to public and private projects (Council of the European Union 1985).[5] Annex I of the directive stipulates projects for which it is mandatory to conduct an EIA, such as railways, roads, waste-disposal installation, and waste-water treatment plants. Member states have discretion over whether to conduct an EIA on projects listed in Annex II, such as some types of agricultural or extractive-industry projects, urban-development projects, and flood-relief projects.

Although the rationale for the EIA directive states that "the effects of a project on the environment must be assessed in order to take account of concerns to protect human health" (Council of the European Union 1985), human health is not explicitly included in the list of direct and indirect effects of a project that must be identified, described, and assessed.[6] Although environmental assessment considers health protection (for example, calculations of safe exposures are included in the derivation of environmental limits for air emissions and water quality), EIAs do not look in detail at the populations likely to be exposed, and compliance with the environmental limits does not mean that there will be no health effects (even small increases in air emissions can have effects on health).

National governments have interpreted the EIA directive differently, and their interpretations determine the extent to which health is explicitly considered in EIA (Bond 2004). For example, the English ministry responsible for planning has resisted including health explicitly in EIA; in contrast, Germany has sought to address health in EIA and passed a resolution in 1992 on HIA in the context of EIA (Fehr et al. 2004). The boundaries are set by bureaucrats in government ministries whose interests often lie in avoiding placement of extra duties on their minister or on businesses. Frequently, the approach taken is to meet legal compliance with minimum expense, and this can result in poor coverage of health. A review of 39 environmental impact statements in the United Kingdom found that 72% did not list human health in the table of contents, 49% provided no analysis of possible human-health effects, and 67% did not include sufficient data to es-

---

[5]The EIA directive has been amended three times (EP/Council 2001, 2003, 2009) to bring it into line with United Nations Economic Commission for Europe Conventions (UNECE 1991, 1998) and to update the list of projects that come under the EIA directive to include those related to transport, capture, and storage of carbon dioxide.

[6]The effects include those on human beings, fauna and flora, soil, water, air, climate and landscape, interaction between them, material assets, and cultural heritage.

timate the number of people potentially affected by the project or activity being considered (British Medical Association 1998 cited in Bond 2004).

A study of the application of the EIA directive concluded that when possible human health effects of a project should be assessed in an EIA rather than by a separate HIA (Hilding-Rydevik et al. 2005). The authors acknowledged that best practice for including health in EIA remains undefined and depends on a number of factors, such as how health is defined (that is, whether it is based on environmental impacts or on a wider array of human health determinants) (Hilding-Rydevik et al. 2005).

In a 2009 survey of the application of the EIA directive, all new member states reported that human health aspects are assessed as part of the EIA reports (COWI 2009). Common elements include the identification of human health effects during the scoping stage of EIA, consultations with health authorities or experts in the field on human health, and assessment of human health effects as a part of the environmental documentation submitted by a developer. Few new member states, however, have produced specific guidance documents for those activities (COWI 2009). Most new member states that were surveyed define health in environmental terms and involve public-health authorities mainly on environmental-health matters. For example, in Hungary, human health issues are examined in the EIA procedures for transport projects (focusing on noise), transmission lines (focusing on nonionizing radiation), hazardous-waste management facilities (focusing on complex effects on environmental health), and strip mines and cement factories (focusing on air pollution). Malta is the only member state that mentions well-being and states that, when relevant, health and well-being are studied with reference to socioeconomic impacts (COWI 2009).

Strategic Environmental Assessment

The SEA directive (EP/Council 2001) refers to public plans and programs but not to policies. The idea is to identify issues at a strategic level so that they do not arise at a project level; in practice, however, the link between strategic assessment and project assessment has proved problematic. Although SEAs are used to evaluate plans in various sectors, they are conducted primarily for land-use planning (EP/Council 2001).[7] If the environmental effects of plans or programs are deemed likely to cross national boundaries, the member state in whose territory the plan or program is being prepared must consult the other member states (EC 2011d). The SEA directive, unlike the EIA directive, explicitly requires the consideration of "the likely significant effects on the environment, including on issues such as ... human health" (EP/Council 2001). The debate on how to include health in SEA is evolving in Europe.

---

[7]SEA is mandatory for plans or programs that are prepared for a prescribed range of sectors and set the framework for granting consent for the future development of projects listed in the EIA directive (EC 2011d).

The SEA directive requires that numerous aspects be examined, including human health, but it does not provide detailed definitions of those aspects. Thus, health is addressed in SEA practice in various ways and in ways that do not systematically require the input of public health or even formal sign-off from health authorities. In Denmark, health is a formal component in the assessment of spatial plans; noise, drinking water, air pollution, recreation and outdoor life, and traffic safety are considered with regard to health (Kørnøv 2009). A review of eight SEAs in England and Germany found that all considered aspects of physical and natural effects (such as noise, emissions, and pollution) on health, and four considered social and behavioral aspects (Fischer 2010). Ensuring that important health effects are satisfactorily identified and considered is challenging, and the SEA directive has not yet led to widespread involvement of public-health experts in the assessment process or in planning. One difficulty is that the health sector tends to be outside the plan-making process, and most HIA experience tends to be at the project level (Cave et al. 2007).

In 2010, the SEA protocol on EIA, which has been adopted by at least 35 member countries, will enter into force (UNECE 2003). It goes much further than the SEA directive in referring explicitly throughout to impacts on environment *and* health, in indicating that all health impacts should be considered (not only those associated with environmental factors), and in indicating that health authorities should be consulted at the different stages of the process.

Examples of Advancing Health Impact Assessment Independently of Environmental Impact Assessment and Strategic Environmental Assessment

Member countries have taken different approaches to advancing HIA outside the environmental-assessment process. Since the late 1990s, Sweden has used HIA as a mechanism for addressing the determinants of health in policy-making (Berensson 2004). Although no legislation requires HIA, agencies, counties, and municipalities continue to learn about and use it. Local politicians across Sweden were actively involved in developing the country's initial guidance documents for HIA and recommended that health be an early part of all policy discussions (SFCC 1998). The decision to screen all political proposals to determine which should be further evaluated led to many policies being recommended for HIA (Nilunger et al. 2003).

Sweden's Health Policy Act of 2003 based its national objectives on health determinants rather than diseases or health problems and linked achievement of the objectives to a monitoring system and annual evaluations. In 2005-2008, 11 central agencies and all of Sweden's county administrative boards were required to implement HIA and were supported by the National Institute of Public Health in doing so (Knutsson and Linell 2010). Although the requirement has heightened interest in and political support for issues related to public health and particularly HIA, there is no legal requirement for HIA, and there are no specific resources for its institutionalization. The result is that implementation is based on the leadership and good will of local individuals and on support by the National Institute of Public Health. Under the existing arrangement, it takes time to develop the capacity for HIA as an integral part of an organization's activities, and work relationships among the sectors have been difficult to achieve (Knutsson and Linell 2010). County administrative boards have made the following observations: legislation or political demand is required to ensure that the public sector implements HIA on a regular basis; the integrated assessment of social, economic, and environmental factors is desirable; and the National Institute of Public Health has used the awareness of EIA processes as a way of structuring HIA approaches and of introducing HIA (Knutsson and Linell 2010). An evaluation of the use of HIA by one health district authority (South West Stockholm) found that a critical factor in the success of HIA was that management at the political and administrative levels had close working relationships, which were achieved and maintained through recurrent opportunities for training and for opportunities in which HIA had the potential to influence policy-making (Berensson 2004).

Slovakia's government passed legislation in 2007 that requires HIA of projects, programs, and policies (O'Mullane 2011). The enforcement of the legislation has been delayed to prepare an institutional framework that includes public-health input. Environmental-health officers in the 37 regional public-health authorities will screen projects to determine which are suitable for HIA. If an HIA is deemed necessary, it will be outsourced to and conducted by the private sector and then evaluated by the regional public-health authority. The National Public Health Authority will have the responsibility of implementing HIA throughout its 37 regional authorities.

Finland has a long-standing interest in incorporating health into all policies and institutionalized HIA for projects in 2002-2006 (Ståhl et al. 2006). The HIAs included stakeholder involvement; were conducted by STAKES, a public-policy institute that has expertise in HIA; and took about 2 months to complete. Local governments were then given the responsibility for HIA and support was provided by the public-policy institute that trained EIA officers and health, education, and local government officials. That process was considered to take too long, and a rapid HIA procedure was developed by STAKES for local government committees to support local-level decision-making. Some cities implemented the procedure successfully, but some sectors objected to impact assessments on particular issues. Where there was resistance to HIA, it was perceived to be a result of the loss of power over decision-making because of the need to consider a wider array of options.

More recently, Finland introduced norms and guidelines for implementing integrated impact assessment, which has been required by law for many years and is led by the Ministry of Internal Affairs. The norms established minimum requirements for impact assessments and allowed questions to be raised if health issues were not included. The Finnish experience points to the essential roles of legislation, of clear process requirements for the implementation of assessments (norms and standards), and of the allocation of budgets (resources) for successful implementation of impact assessment. In the Finnish context, the use of integrated impact assessment, which is required by law, is seen as the best way to integrate health and environment issues into policy-making (T. Ståhl, National Institute of Public Health, Tampere, Finland, personal communication, 2011).

In England, an act of Parliament stipulates that all strategies passed by the mayor of London must reduce health disparities in London (HM Government of Great Britain 2007). That requirement means that public-health input is required in all sectors, it places health assessment firmly within the policy process, and it makes the reduction of health disparities a matter that spans the activities of the Greater London Authority. London strategies for transportation, housing, employment, and education have all been subject to HIA (Opinion Leader Research 2003; Bowen 2004; London Health Commission 2011), and capacity for HIA has been developed at regional and local levels around London.

Alternative approaches for advancing HIA at local levels have focused on particular funding programs. For example, in 2000, a redevelopment program in Wales required that all proposals take health into account; accordingly, HIAs had to be completed to ensure that proposals were funded (see, for example, Breeze and Kemm 2000). The Welsh Assembly has formed a special unit to assess health impacts of proposed legislation and advise parliamentarians (Breeze and Kemm 2000).

National and local requirements for HIA may be supported by information repositories, for example, the HIA gateway,[8] which was funded by the English Department of Health. Advisory bodies have also supported and propagated the use of HIA. For example, the Welsh Health Impact Assessment Support Unit at the Cardiff University School of Social Sciences was formed in 2001. It formed a partnership with the National Public Health Service and works to develop capacity for HIA in Wales, provide information and advice, and conduct research and evaluation. Examples of similar centers in other parts of Europe include the National Institute for Public Health and the Environment (RIVM) in the Netherlands; the Institute of Public Health, North Rhine Westphalia, in Germany; and the Unit for Health Promotion Research in the University of Southern Denmark. Some centers have informal oversight and advisory roles. RIVM, for example, provides policy advice to the Ministry of Health in environmental health and chronic diseases and specializes in the quantification of health effects in which HIA expertise plays a role (L. den Broeder, National Institute for Public Health and the Environment [RIVM], the Netherlands, personal communication, 2010).

The experience of the World Health Organization (WHO) Europe Healthy Cities Network (HCN) provides some lessons that could be instructive for the United States by suggesting how U.S. cities and counties might adopt and adapt HIA. It also indicates the magnitude of the work and time required to achieve the change in policy infrastructure to advance HIA and to ensure that all sectors are comfortable and confident with the process. The HCN is made up of more

---

[8]See www.hiagateway.org.uk.

than 90 European cities (WHO 2011). To join the HCN, cities must apply, they must fund their input, and they must demonstrate a high level of political support. Thus, the cities involved are, in theory, willing partners and are keen to learn from the HCN and adapt their policies accordingly. Since 1998, healthy urban planning has been a part of the network (Barton et al. 2009), and capacity-building and peer support have always been important elements of the movement as a whole. Phase IV of the WHO Healthy Cities Project ran from 2003 to 2008 and included HIA as one of its core activities. In 2003, the focus was on adoption of HIA; the two main types of barriers to adoption were characterised as technical and political (Ison 2009). Suggestions for overcoming technical barriers included providing training; technical support, particularly in initial HIAs; mentoring; and peer review. Suggestions for overcoming political barriers included "increasing political understanding of what HIA is and what it can offer; involving the politicians at a strategic level in setting the conditions for the use of HIA by the municipality; piloting HIA with proposals that are likely candidates to increase the potential for health gain; [and] presenting the results of HIA in a useful and useable format for politicians so that health can be taken into consideration during decision-making" (Ison 2009, p. i69). Internal evaluation of the HCN found that it has advanced HIA in several municipalities in the region and has sensitized other municipalities to the relevance of creating health gain (Ison 2009); however, there is no consideration of the effectiveness of the HIAs that the HCN recommends.

The European experience has shown that capacity-building is important, particularly for knowledge transfer within and between organizations. Although it is recognized as a useful approach, HIA is rarely a core responsibility listed in job descriptions. Public-health specialists in the health sector find it difficult to dedicate time to HIA, and there is no clear career path for young professionals who wish to pursue HIA. In the United Kingdom and Ireland, there are a few short courses in HIA that are seen as part of continuing professional development. At the University of the West of England, a substantial proportion of planners are trained in the Faculty of the Built Environment. The university has a large public-health school and requires planners and public-health professionals to take a course in each other's field. In some respects, the lack of capacity is being met by the private sector as specialists in environmental assessment are starting to add HIA to their skill set. Although working across sectors is desirable, increasing the capacity for HIA outside the discipline of public health will have long-term implications for the development of HIA.

AUSTRALIA

This section reviews the Australian experience in addressing health in EIA, in advancing HIA by using alternative methods, and in strengthening the consideration of health equity in HIA.

Health in Environmental Impact Assessment

EIA was established in Australia in 1974 by the Environmental Protection (Impact of Proposals) Act, which is applicable to national-level decisions and includes provisions for assessing vectorborne diseases associated with the construction of large dams (Australian Government 2009). In 1994, the National Health and Medical Research Council—a research body tasked with improving public health in Australia—published findings on health in EIA (NHMRC 1994). They found a lack of structures and processes for incorporating health in EIA, inconsistent coverage of health in EIA because of gaps in EIA legislation, and inadequate involvement of health agencies in the EIA process. The council proposed integrating HIA into EIA and stated that "human health is affected by social, psychological, economic, ecological, and physical factors" (NHMRC 1994, p. xii). It defined a framework for integrating health into EIA, specified ways to access public-health expertise and to finance community involvement, and identified methodologic issues to be addressed (Harris and Spickett 2011).

The Australian federal government established the enHealth Council, a national body with responsibility for implementing a National Environmental Health Strategy and providing leadership on integrating health into EIA (Harris and Spickett 2011). The council published guidelines for implementing HIA (enHealth Council 2001) that promoted the integration of health into EIA, the consideration of the social determinants of health, and the recognition of the broader application of HIA beyond projects and into policy and program development. Furthermore, the guidance emphasized the need to assess adverse *and* beneficial health effects and overcome the previous tendencies in EIA to assess only adverse effects. In 2005, however, an analysis found that legislative and administrative frameworks and procedures needed for facilitating HIA implementation were still lacking (NPHP 2005). The responsibility for HIA was later defined to be a matter of state and local jurisdiction, not a federal-government responsibility.

At the state level, Tasmania introduced legislative requirements in 1996 for the conduct of HIA as part of the EIA process (Government of Tasmania 1994). That legislation was one of the first examples of requirements for the consideration of health effects (in addition to environmental effects) to be formally legislated. Mahoney (2009), however, suggested that the requirements were due more to the configuration of the government department responsible for public health and environmental management than to a calculated decision to set priorities related to health. Queensland is the only other Australian state that combines health and environmental effects, and it too has been successful in addressing health in EIA (Mahoney 2009).

In 1998, Tasmania published a manual to guide local governments in the execution of their public-health and environmental-health duties (Public and Environmental Health Service 1998). It emphasized environmental risks to health, the monitoring of those risks, and detailed risk-assessment methods. However, HIA was ultimately impeded because of a lack of sufficient workforce capacity. Later revisions of HIA procedures encouraged more efficient scoping and earlier interactions between developers and appropriate government agencies (Harris and Spickett 2011).

Health Impact Assessment Independent of Environmental Impact Assessment

Around 2000, there was a move to promote HIA independently of EIA as a way to influence healthy public policy (such as in the transportation and housing sectors) and to place an emphasis on stakeholder participation in the HIA process and on the social determinants of health (Mahoney and Durham 2002). A federal research-grant program supported that work, including communication among sectors, tools development, and capacity-building. The grant program was disbanded after a few years, but an effort in capacity-building for conducting HIAs of policies continued in two states. In New South Wales, the focus was to build health-system capacity to implement HIA. In Victoria, the focus was on local government planning systems.

Equity-Focused Health Impact Assessment

Equity-focused HIA and a framework for considering the differential distribution of impacts explicitly were also developed in Australia to support Health In All Policies (Mahoney et al. 2004). The framework was tested in case studies and succeeded in placing the focus of HIA on the equitable distribution of health in the population (Simpson et al. 2005). New South Wales included HIA in its strategy to reduce health inequity and funded HIA capacity-building, and the Centre for Health Equity Training, Research and Evaluation of the University of New South Wales has conducted rapid equity-focused HIAs.

Advancing Health Impact Assessment in Australia

The development of HIA throughout Australia has been influenced by priorities at the state and territory levels (Harris and Spickett 2011). For example, Western Australia and the Northern Territory have active mining sectors; environmental assessments of many of the mining activities are conducted, and HIA is integrated into them. With support from the University of New South Wales, the states of New South Wales, Queensland, South Australia, and Victoria are building capacity for and supporting the implementation of HIA as a tool for healthy public policies through learning-by-doing programs (that is, programs that emphasize learning through participation).

The Australian experience indicates that "system support and capacity-building" may do more to promote HIA than legislating its use (Harris and Spickett 2011). New South Wales, Western Australia, and South Australia do not have any legislative requirements for HIA but have advanced HIA with some support of the health sector and others across the government. Although legislation requires HIA to be included in EIAs in Tasmania, there have been difficulties in applying HIA within a regulatory process of a nonhealth agency, including lack of sufficient workforce capacity and efficient procedures for communicating between proponents and relevant agencies. Victoria has incorporated HIA into its Public Health and Wellbeing Act 2008 by systematically investing in the positioning of HIA in local government as a tool for healthy public policies and by building capacity for HIA among public-health staff.

THAILAND

Over the last decade, Thailand has developed a comprehensive system for HIA. The National Health System Reform (NHSR) was launched in 2000 and has advocated for addressing health in policies in nonhealth sectors and for a greater role for the public in decision-making. HIA was identified as a mechanism for developing a healthier society by facilitating stakeholder involvement and by including sound information in public policy-making (Phoolcharoen et al. 2003).

Public policies to transform Thailand into an industrialized economy were met with civil unrest and set the historical background for the move to increase the public's role in decision-making. The 1997 constitution created mechanisms for participatory decision-making, resource allocation, decentralization, greater accountability, and transparency. The NHSR process reflected the national objectives, and in 2001, an NHSR commission funded research to inform the National Health Act and to develop HIA in Thailand. The first attempt to introduce HIA into the EIA process was not successful. It was concluded that EIA would need to be modified to allow for broader participation. Moreover, at that time, knowledge of, experience in, and skills for HIA were lacking in Thailand. The low level of capacity was identified as a threat to the credibility of HIA if it were to develop as a formal approval mechanism. In 2002, the Ministry of Public Health established a Division of Sanitation and Health Impact Assessment to define HIA systems and to support healthy public policy, especially among local governments. The focus changed in 2003 to HIA in healthy public policy as a learning process, and this process was to be developed in parallel with obtaining support for the concept of the NHSR and with development of a critical mass of HIA knowledge and skill in the country.

In 2005, the National Economic and Social Advisory Council—which had experience with implementing HIA in a variety of projects and policies—submitted HIA recommendations to Thailand's cabinet. The recommendations were accepted, and the Ministry of Health was directed to implement them. A clear mandate for HIA in Thailand was established as a way to stimulate greater interest in developing healthy public policy. The 2007 federal constitution requires EIA and HIA and states that a public hearing must take place to obtain the opinion of interested parties and others who might be affected by a project or activity and that a community has the right to sue any government agency that does not comply (Thai Laws 2007). The National Health Act, also issued in 2007, includes the right of people to ask for and participate in an HIA of a public policy, and it requires the NHSR Commission to develop HIA guidelines and procedures (NHC Thailand 2007). The National Development Plan of 2007-2011 includes provisions for "integration of health in the EIA system" and "applications of SEA...with health considerations in main sections and in spatial planning" (NHC Thailand 2008).

Experience with implementing HIA continues to evolve. Some of the successes include the integration of health assemblies with multistakeholder participation into policy-making at a local level. HIAs have been used for healthy agriculture policies at local and regional levels, industrial policies, and water management. HIA credibility has been found to depend on who conducts the assessment; HIAs led by a health department using participatory learning have had better results than HIAs led by nongovernment organizations and local experts. HIA recommendations that require changes in existing business practices have needed substantive analysis to support them. For example, HIAs evaluating policies for the production of healthier foods tried to demonstrate the relative health costs of current business practices compared with those of policy alternatives *and* to develop measures that would influence consumer demand for healthy foods (Elinder et al. 2003; Cole et al. 2007).

Some of the present challenges for Thailand are to define specific mechanisms for public participation and for incorporating the results of the assessment into policies; to develop rules, regulations, and guidelines for HIA in specific sectors, such as agriculture and food production, that take into account sector issues and business-management practices; to expand the knowledge base so that the health burden of policies and methods can be recognized; and to identify short-term and long-term effects of the policy.

WORLD HEALTH ORGANIZATION

HIA initially developed in WHO in response to the need to control vector-borne diseases resulting from water projects without using chemicals. In 1981, the Panel of Experts on Environmental Management for Vector Control was established to develop institutional frameworks for intersectoral and interagency collaboration. The panel developed methods to forecast diseases in water-management projects (Birley 1991). Further development of HIA occurred when the World Commission on Dams, a multistakeholder commission, made recommendations for the sustainability of dams and set good practice standards. The commission expressed concern over health impacts (Colson 1971) and, in cooperation with WHO, included health assessments in its deliberations (WHO 2000a).

WHO's Regional Office for Europe (EURO) supported training for the inclusion of health in EIAs beginning in the 1980s (Tiffen 1989; WHO 2000b). Other WHO regional offices—such as the Regional Office for the Eastern Mediterranean (Hassan et al. 2005) and the Pan American Health Organization (Weitzenfeld 1996)—prepared HIA guidelines that focused on addressing environmental determinants of health; some of the guidelines have been widely used. An HIA training package with guidance for government cross-sector policy-making was issued in 1999 and implemented in several countries (WHO 2000b). In 2003, an HIA Web site[9] was established, and a special-themed issue of the *Bulletin of the World Health Organization* was dedicated to HIA experience at that time (Volume 81, Number 6).

Since the late 1990s, WHO's focus on HIA has included applications in industrialized countries. The focus on HIA broadened from incorporating health into EIAs to developing healthy public policies. WHO EURO supported HIA in specific sectors, including agriculture (Lock et al. 2003) and transportation (Dora and Racioppi 2003). WHO EURO also developed a project to learn from HIA experience and clarify basic concepts and definitions, principles, approaches, and methods used in HIA. A series of reviews and meetings were carried out, and support for continued learning was provided through a network to decision-makers. Those activities led to the publication of the Gothenburg consensus paper (Diwan et al. 2000). Also, as previously discussed, a project on HIA was developed by the Europe Healthy Cities Network (EU 2009).

The WHO experience with EIA and HIA for healthy public policy was used to inform and influence the negotiations of the new SEA protocol to the United Nations Economic Commission for Europe Convention on EIAs (Dora 2004). The final text included a broad health perspective, placed health as a key aspect of the SEA, and specified ways to include health (UNECE 2003). That same broad perspective was successfully used in a project to support healthy public policies, including the use of HIAs in Uganda, Jordan, and Thailand that focus on agriculture, livestock, and water-management policies.[10] WHO EURO has also assisted several countries in its region in conducting HIAs of climate change, and a few countries have developed national adaptation plans that include specific consideration of health (WHO 2008).

Tools for HIA oversight have recently been developed by WHO to be used by multilateral development banks and recipient countries. Those tools support the inclusion of health goals in development lending for all sectors of the economy (M. Pfeiffer and C. Dora, WHO, unpublished material, 2010), and they support a decision by the International Finance Corporation (IFC) to adopt safeguards for community health and safety. Integrating health into development lending through the use of HIA has the potential to influence large public and private-sector investments in developing countries, including natural-resource extraction (such as oil, mining, and forestry), infrastructure, and tourism. WHO

---

[9]See http://www.who.int/hia/en/.
[10]See http://www.who.int/heli/pilots/en/.

is working with a few pilot countries on the development of governance mechanisms in the extractive industry for healthy public policy by including HIA and connecting health with national planning processes.

In 2008, the Commission on Social Determinants of Health (CSDH) recommended that WHO support health-equity impact assessments of important global, regional, and bilateral economic agreements and in all government policies, including finance, as a way to address health disparities. It was recommended that member states of WHO redesign their health sectors to integrate a focus on social determinants of health into relevant sectors (CSDH 2008). To achieve that goal, WHO proposed that countries adopt and perform HIAs for policies and projects and focus further on health equity. The CSDH also recommended that data systems present information disaggregated by sex, socioeconomic status, and other criteria to allow for the identification of disparities; it warned that public participation does not necessarily ensure that equity issues are addressed; and it called for capacity-building to assess the health-equity impacts of major global, regional, and bilateral economic agreements and to monitor social determinants of health and health equity.

MULTILATERAL DEVELOPMENT BANKS

Multilateral development banks provide financial support and advice for economic and social activities in developing countries. They include the World Bank and associated institutions—the African Development Bank, the Asian Development Bank, the European Bank for Reconstruction and Development, and the Inter-American Development Bank Group. The World Bank expects its borrowers "to integrate selected environmental and social aspects into the identification, planning, appraisal, and implementation of the investment projects that it supports" (Mercier 2003, p. 461). To facilitate compliance, the bank requires a series of impact assessments, including assessments of projects that might have effects on the environment, natural habitats, forests, safety of dams, pest management, indigenous peoples, involuntary resettlement, cultural property, and international waterways and projects in disputed areas (World Bank 2011). Such assessments are considered safeguards and were adopted gradually between 1998 and 2006. The assessment reports must be disclosed in the countries in which the projects are expected to be implemented and on the World Bank InfoShop Web site, and they are expected to be communicated internationally. The World Bank does not have a safeguard for public or community health. It has defined EIA to include the natural environment, human health, safety, and social aspects; in 1996, it commissioned guidance for including health as part of an EIA (Birley et al. 1997).

The Asian Development Bank has had a concern about the health consequences of projects that it funds and commissioned guidelines in 1992 for the HIA of development projects (Birley and Peralta 1992). The guidelines are brief and have not been updated recently, but the bank has had a staff member who has HIA expertise on its safeguards team for many years.

The IFC—the private-sector lending arm of the World Bank—lends to private-sector investors primarily in developing countries for such projects as the extractive industry or tourism. The IFC adopted safeguards for projects submitted for funding and developed a set of criteria for assessing potential impacts on the environment, employment, occupational health, and safety.

In 2006, the IFC developed additional safeguards for public health after a debate about the oversight of adverse health impacts of projects funded by IFC that could possibly pose a risk to businesses and therefore to the IFC itself (IFC 2006). The new safeguards, referred to as performance standards, added a standard on community health and safety to several existing standards on occupational health and safety, and IFC produced guidelines and a benchmark for industry to help it meet the new standards (IFC 2007). In 2009, the IFC published guidance for carrying out HIAs that covered potential health issues in large-scale projects in developing countries in, for example, the extractive industry (IFC 2009). That requirement is potentially beneficial for public health because all projects for which the IFC is one of the co-financers will need to have the community-health and safety-assessment performance standards included. Private investment is a large fraction of the financial investment in developing countries today.

In 2003, the Equator Principles Financial Institutions (EPFIs)—a group of 67 private banks, including some in developing countries—agreed that no loans should be provided to applicants that would not or could not comply with social and environmental policies and procedures modeled after the environmental standards of the World Bank and the social policies of the IFC. The EPFIs have only recently been trained to implement the new IFC performance standards, and there is no independent mechanism for assessing compliance or quality assurance in the implementation of the standards. A network of professionals are engaged in the implementation of the performance standards in those banks in an effort to facilitate learning from experience. The Equator Principles have become the voluntary standards for private banks in assessing development projects. Those measures have the potential to include health and other criteria in private-sector lending. Accountability mechanisms will need to be built into the system at some point and could provide the incentive for better performance, as shareholders and other interested groups identify the actual contributions of private bank lending to promoting health and other development criteria.

UNITED STATES

The first use of a process identified as HIA in the United States occurred in 1999 in the context of a policy to increase the minimum wage for San Francisco contractors and leaseholders (Bhatia and Katz 2001). That use of HIA contributed to the passage of an ordinance and an increase in the minimum wage

(Dannenberg et al. 2008). The early use of HIA by a U.S. government agency was focused on the integration of public-health agency expertise into local land-use planning decisions principally in the San Francisco Bay area of California. The use of HIA then began spreading to other parts of the country as an independent practice with some expansion of the breadth of policy sectors and more recently as an enhancement of the health analysis conducted in the state and federal systems for EIA. In 2010, there were a growing number of examples of the use of HIA in the United States in a wide variety of agencies at the local, state, and national levels.

Local Communities

The use of HIA in local communities has spread substantially over the last decade. Surveys in 2010 show HIA being used in a number of large metropolitan areas and medium-size communities for a variety of actions (Dannenberg et al. 2008; UCLA HIA-CLIC 2011). HIA of policies, projects, and programs in local communities has been organized or sponsored by local public-health agencies, nonprofit organizations, planning agencies, and academic institutions. For example, several HIAs focus on individual development projects or community plans (Farhang and Bhatia 2007; Heller et al. 2009; Human Impact Partners 2009) whereas the BeltLine HIA evaluated a regional redevelopment and transportation project in the greater Atlanta metropolitan area (Ross 2007). Land-use, housing, and transportation planning have been more common foci of HIA than policies or programs in the labor, education, or social-services sectors. In the transportation context, current HIA work includes analysis of transportation infrastructure proposals in the Minneapolis-St. Paul area, the Houston Urban Corridor, and the Los Angeles area and a proposed road-pricing policy in San Francisco (UCLA PH 2011; ISAIAH 2011; SFDPH 2011; UCLA HIA-CLIC 2011).

HIA has also been used to gauge the health impacts of proposed changes in local zoning ordinances. The Eastern Neighborhoods Community Health Impact Assessment, completed in 2006, analyzed three rezoning plans for former industrial neighborhoods and focused on issues of displacement and environmental quality (Corburn and Bhatia 2007; Farhang et al. 2008). Another recent example is from the city of Baltimore, where an HIA found that the city's proposed zoning code would have several implications for health. The HIA team noted that "if implemented, the draft new code could substantially increase the percentage of residents who live in neighborhoods that allow mixed use. This has the potential to increase residents' physical activity levels as well as access to healthy food. [It could also] dramatically increase the percentage of neighborhoods that allow urban gardens and farmers markets. This has the potential to increase residents' access to healthy food if these uses were developed" (Thornton et al. 2010, p. 1-3). The HIA made several recommendations for modifying the zoning code to promote health.

Although public-health agencies in several localities—including Denver, Baltimore, Seattle, Portland, Los Angeles, and the North Slope Borough in Alaska—have been either leaders or participants in HIA initiatives, it is less common for public-health agencies to have incorporated HIA as a routine day-to-day institutional practice. Over the last decade, however, the use of HIA in San Francisco has matured to become an integral part of the work of the Department of Public Health with dedicated public funding and staff since 2002. HIA tools are now routinely applied in partnership with other city agencies, including planning and redevelopment agencies, to evaluate such proposals as neighborhood and community plans. The San Francisco Department of Public Health has established a routine role of providing oversight of environmental health analysis in EIAs implemented under the California Environmental Quality Act. It has also been involved in the institutionalization of HIA practice through training and evaluation partnerships with the University of California, Berkley and research initiatives to develop analytic tools and approaches to address methodologic gaps. Several HIAs conducted by the San Francisco Department of Public Health have been implemented in close partnership with or under the oversight of nongovernment organizations—a fact that may be instrumental in the continuing community demand for HIA (Corburn 2009). Notably, community demand is leading to a broadening of the scope of practice beyond physical planning to policies related to labor rights and working conditions.

States

In 2006, Washington became the first state to pass legislation focused on enabling preparation of health impact reviews. A health impact review has been defined as a "review of a legislative or budgetary proposal…that determines the extent to which the proposal improves or exacerbates health disparities" (Revised Washington Statutes 43.20.015). The state legislature made formal findings that women and people of color experience important disparities compared with men and the general population and that the disparities affect health in many ways. The state also expressed an intent "to create the healthiest state in the nation." The law established a mechanism under the purview of the State Board of Health to undertake health impact reviews on the request of a state legislator or the governor (Revised Washington Statutes 43.20.285). Although some reviews have been requested (WA SBOH 2007a,b; 2008a,b,c; 2009, 2010), state budget difficulties have resulted in diminished capacity to conduct reviews.

Massachusetts has also passed legislation to support HIA, and bills to support HIA have been proposed in California, Maryland, Minnesota, and West Virginia. Most of the bills would provide for an expanded role for state health agencies in HIA and related planning efforts. However, even without the incentive of legislation, some state health departments have become more engaged in HIA over the last few years. To date, state-level administrative actions include the establishment of interagency working groups and pilot programs and technical assistance to agencies that are developing regulations that may affect public health. For example, the Hawaii Department of Agriculture is partnering with the Kaiser Permanente Center for Health Research and the Kohala Center—a nonprofit organization focused on community education, research, and conservation—to develop an HIA that will inform the development of a Hawaii County Agriculture Development Plan. The plan is being developed in the wake of the demise of the sugar plantations that used to dominate the agricultural economy on the "big island" of Hawaii and the disappearance of many smaller agricultural producers (UCLA HIA-CLIC 2011). In Alaska, the Department of Health and Social Services has established an HIA program to provide technical assistance to other agencies involved in conducting integrated environmental and health impact assessments (Alaska HSS 2011).

In California, the Department of Public Health recently became the first state agency to publish an official guidance document on HIA (Bhatia 2010). In 2009, the California Air Resources Board, in partnership with California Department of Public Health, initiated an HIA of proposed cap-and-trade regulations required to be promulgated under the California 2006 Global Warming Solutions Act. The act directed the California Air Resources Board to adopt regulations that avoid, to the extent feasible, disproportionate impacts on low-income communities (California State Health and Safety Code, Division 25.5, Greenhouse Gas Emissions Reductions, § 38562(b) (2)). The act also mandated that, in the development of the regulations, consideration be given to overall societal benefits, including public health. A second phase of the HIA was recently initiated to expand the scope of the analysis with external funding and support by the private nonprofit Public Health Institute. Other state health departments engaged in HIA include those of Wisconsin, Oregon, Washington, Massachusetts, and Alaska (Cagle 2010; WI DPH 2010; ANTHC 2011; Oregon Government 2011).

Federal Government

The use of HIA in decision-making at the level of the federal government has been largely, although not exclusively, in the context of implementing NEPA. Federal agencies in the executive branch of government have been, in theory, required to assess the health effects of proposed federal actions under NEPA since its passage in 1969 (42 U.S.C. §§ 4321-4347).[11] The language in NEPA that embodies the threshold for the preparation of an environmental impact statement (EIS) uses the phrase "the quality of the human environment" (Congressional Record, Senate, P. 40416, December 20, 1969) because the congressional sponsors intended to demonstrate that "an environmental policy is a

---

[11]There are some gaps in coverage under the statute, most notably for these purposes the pollution-control regulatory activities of the U.S. Environmental Protection Agency.

policy for the people. Its primary concern is with man and his future" (Congressional Record, Senate, p. 40416, December 20, 1969). Indeed, the statutory purpose of NEPA includes promoting the "*health* and welfare of man" (42 U.S.C. § 4321; emphasis added), and the national environmental policy—whose articulation and implementation were the major purpose of the act—includes assurance that all Americans are entitled to "safe, healthful, productive, and aesthetically and culturally pleasing surroundings" (42 U.S.C. § 4331) and the attainment of the "widest range of beneficial uses of the environment without degradation, risk to *health* or safety, or other undesirable and unintended consequences" (42 U.S.C. § 4331; emphasis added).

Similarly, the regulations implementing the procedural provisions of NEPA include health as an important focus of analysis.[12] The direct, indirect, and cumulative health effects of proposed federal actions are to be analyzed under NEPA (40 C.F.R. § 1508.8), and the degree to which a proposed action affects public health or safety is one of the criteria for determining whether preparation of an EIS is required (40 C.F.R. § 1508.27). Furthermore, Congress directed the administrator of the U.S. Environmental Protection Agency (EPA) to review and comment in writing on the analysis of the impacts of proposed actions. EPA was asked to refer any proposed legislation, action, or regulation that fell under the auspices of NEPA and that was determined to be "unsatisfactory from the standpoint of public health or welfare or environmental quality" (42 U.S.C. § 7609 [1970]) to the Council on Environmental Quality (CEQ), which is the environmental agency in the executive office of the president.

From a procedural perspective, there is no significant difference between the steps in HIA and EIA, at least as practiced under the regulations implementing NEPA. Both processes begin with the identification of proposed actions that should go through the process, as opposed to proposed actions that are likely to cause no or *de minimis* impacts. In HIA, this step is called screening (described in detail in Chapter 3). Under NEPA, agencies are required to publish procedures that provide categories of actions that an agency has determined generally require the preparation of EISs and environmental assessments and actions that are excluded from written documentation.

The next step for both processes is a period of scoping to identify important issues, interested parties, and work that needs to be done to prepare a credible analysis. The analysis itself is subject to public review and input and includes mitigation measures and alternative ways of achieving the goal. Under NEPA, agencies are required to disclose their decision about a proposed action that is subject to an EIS in a "record of decision" (40 C.F.R. § 1505.2). HIA

---

[12]Government-wide NEPA regulations binding on all executive branch agencies were promulgated by the Council on Environmental Quality, an agency established by Congress under NEPA in 1979 to, among other things, advise the president on environmental matters and oversee implementation of NEPA (40 C.F.R. §§ 1500-1508). The statute, regulations, and other useful reference material can be found at www.nepa.gov.

does not have codified requirements, but the intent is to have HIA considered in the course of decision-making.

Despite the clear emphasis on analysis of human health impacts and concern for public health as a primary element of the quality of the human environment, several factors have led to a historical tendency to minimize the importance of human health effects in the context of NEPA analysis. Litigation has had a major influence on the shape, development, and perception of NEPA law, and specific claims related to human health were seldom a major early focus. Some confusion stemmed from several early NEPA cases that held that social and economic effects themselves did not trigger the requirement to prepare an EIS (although health effects were not the subjects of claims in the cases). For example, residents living near the Three Mile Island nuclear power plant had fears related to the restart of the plant after a partial core meltdown. A decision was made by the U.S. Supreme Court that the fears did not need to be analyzed by the Nuclear Regulatory Commission under NEPA, and this was interpreted by some to mean that health effects were not subject to challenge under NEPA. That interpretation is wrong; indeed, all members of the U.S. Supreme Court concurred in the statement and were of the opinion that "all the parties agree that effects on human health can be cognizable under NEPA, and that human health may include psychological health" (*Metropolitan Edison v. People Against Nuclear Energy*, 460 U.S. 766, 771 [1983]).

Another factor that has led to the minimization of human health effects under NEPA is that the federal agencies that have been the focal point of activist, legal, and legislative attention in the NEPA context tend to be agencies that traditionally have not had internal expertise in matters of public health (for example, the U.S. Army Corps of Engineers, the Forest Service, and the Federal Highway Administration). In contrast, federal agencies whose mission is focused on health—such as the U.S. Centers for Disease Control and Prevention (CDC)—have seldom been the focus of attention from a NEPA perspective. Professional and functional collaboration between the two sets of federal institutions in the context of NEPA has, until quite recently, been unknown.

The confusion generated by misinterpretations of case law, the separation of agency cultures and professional exchanges, and a lack of vigorous advocacy have resulted for several decades in unintended sidelining (although not complete omission) of analysis of health effects in the context of the NEPA process, which includes the analytic and procedural EIA processes under NEPA. The situation began to change as the concept of HIA was introduced into federal agencies. Native Alaskan villagers had long-standing concerns about the impact of oil and gas leasing on subsistence hunting and fishing and the associated health, social, and cultural impacts. Their concerns began to receive attention from federal agencies when work was initiated on behalf of Native Alaskans to introduce the concept of HIA into those agencies (see Box 3-3 in Chapter 3). It was shown that HIA was modeled after and easily integrated into EIA under NEPA, and professional assistance was provided to interested parties. The result was that a health-effects analysis was included in several NEPA documents for oil and gas leasing programs and lease sales (BLM 2007; MMS 2007a,b; EPA 2009). Publication of the documents sparked attention and interest in other agencies. For example, the CEQ hosted a presentation about HIA for federal agency personnel who work on the implementation of NEPA (H. Greczmiel, Council on Environmental Quality, Washington, D.C., personal communication, 2010). EPA recently supported a model scoping exercise for HIA of future port expansion projects in Los Angeles, which generally also require environmental review (EPA 2010). CDC and EPA signed a memorandum of understanding in 2002 to collaborate and strengthen the understanding of linkages between proposed changes in the built and natural environment and potential health outcomes, a step that should be of benefit to many agencies in the context of the NEPA process.

Public-interest organizations have also become more aware of HIA and have been advocating, with mixed success, its inclusion into a wider array of NEPA analyses. The Natural Resources Defense Council, for example, now advocates the inclusion of a comprehensive assessment of potential human health impacts in EISs that analyze the impacts of oil and gas exploration and production on federal lands (Mall et al. 2007). In another example, a broad coalition of community interests and government representatives has asked that an HIA be conducted on the expansion of the I-710 freeway in Los Angeles County—a project undergoing environmental review under NEPA and the California Environmental Quality Act (Los Angeles County Metropolitan Transportation Authority 2010).

Other federal authorities also call for an assessment of health risks. Executive Order 12898 (Federal Actions to Address Environmental Justice in Minority Populations and Low-Income Populations) reinforces the inclusion of a systematic analysis of health issues in NEPA documents by instituting a requirement that agencies recognize and address the "disproportionately high and adverse human health or environmental effects" of federal actions on low-income and ethnic-minority populations (EO 12898, 59 Fed. Reg. 7629 (Feb. 16, 1994)). In essence, that executive order creates a two-step requirement in which agencies must first identify potential adverse health effects of agency actions and then determine whether the effects are likely to affect low-income or minority populations disproportionately. The order thus reinforces the basic NEPA requirements regarding health but further recognizes that in some cases ethnic-minority and low-income populations may be more vulnerable to adverse health effects of agency decision-making.

The CEQ (1997, p. 9) issued detailed guidance on the implementation of Executive Order 12898 and in it advised agencies to

> consider relevant public health data and industry data concerning the potential for multiple or cumulative exposures to human health or environmental hazards in the affected population and historical patterns of exposure to environmental hazards, to the extent such information is reasonably available. For example, data may suggest there are dispropor-

tionately high and adverse human health or environmental effects on a minority population, low-income population or Indian tribe from the agency action. Agencies should consider these multiple, or cumulative effects, even if certain effects are not within the control or subject to the discretion of the agency proposing the action.

It should be noted that agencies under NEPA are required to analyze effects, whether they are within the control and responsibility of the proponent agency or not. The issue of what agencies can require outside applicants to carry out in the way of mitigation measures is less clear if a mitigation measure in question involves actions arguably outside an agency's jurisdiction (*Cape May Greene, Inc. v. Warren*, 698 F.2d 179 (ed Cir. 1983)). Furthermore, although NEPA requires the analysis of mitigation measures, the U.S. Supreme Court has ruled that NEPA does not require agencies to adopt any particular mitigation measures (*Robertson v. Methow Valley Citizens Council*, 490 U.S.332 (1989)).

Executive Order 13045 created similar requirements for agencies to identify and address actions that could have disproportionate effects on children: each federal agency "(a) shall make it a high priority to identify and assess environmental health risks and safety risks that may disproportionately affect children; and (b) shall ensure that its policies, program activities, and standards address disproportionate risks to children that result from environmental health risks or safety risks"(EO 13045, 62 Fed. Reg. 19883 [April 23, 1997]).

Health Impact Assessment Independent of the National Environmental Policy Act

The practice of HIA has also been used for federal decision-making outside the NEPA process. For example, one HIA analyzed the effects of the Healthy Families Act in 2008, and legislation was proposed that mandated 7 sick days a year for businesses that have more than 15 employees (Bhatia et al. 2008). The HIA and additional research for similar legislation in Massachusetts found potentially substantial health benefits and cost savings resulting from such legislation. Those HIAs were considered by members of Congress when the proposed legislation was discussed.

Another example is a rapid HIA that was prepared as a demonstration project for the 2002 Federal Farm Bill by the School of Public Health HIA Project at the University of California, Los Angeles (UCLA) (UCLA PH 2004). The analysis identified major pathways through which the bill could affect health and focused on two of them (dietary consumption and air pollution). Data limitations prevented analysis of the other three pathways (food safety, rural income and quality of life, and environmental degradation).

Other recent developments set the stage for further consideration of HIA at the federal level. First, the Affordable Care Act of 2010 (Pub. L. 111-114) calls for a National Council on Prevention, Health Promotion, and Public

Health. Established by President Obama in June 2010 (EO 13544 [June 10, 2010]), the council is composed of cabinet-level and other senior administration officials in both health and nonhealth agencies and is chaired by the U.S. surgeon general. The council's mission is to examine the interplay of factors that affect public health. Among provisions laid out by the council's framework for the National Prevention Strategy is a call for a cohesive federal response to prevention, for a reduction of health disparities, and for support of healthy physical and social environments (NPHPPHC 2010). This focus, with requirements for annual reports from the council, will help to sustain attention to the multiple determinants of health and related improvement opportunities, such as HIA. Second, the Healthy People 2020 program of the U.S. Department of Health and Social Services establishes national goals and objectives for addressing the major health challenges in the United States.[13] The current version of the program includes an expanded focus on the social determinants of health, and the Secretary's Advisory Committee on National Health Promotion and Disease Prevention Objectives for 2020 discusses the use of HIA to achieve the program's objectives. Third, the White House Task Force on Childhood Obesity issued a report to President Obama in May 2010. The report encourages communities to consider the impacts of built-environment policies and regulations on human health and to consider integrating HIA into local decision-making processes (White House Task Force on Childhood Obesity 2010).

Academic Institutions

Several academic institutions have helped to advance HIA. The University of California, Berkeley has offered a master's-level course in HIA. The course is designed as project-based learning, and students complete full-scale HIAs of contemporary decisions of local, regional, or state significance. In several cases, the HIAs produced by students in the course have been used by local community organizations or public agencies to inform decision-making (UCB HIA 2011). The student HIAs have both demonstrated the innovative use of research methods that have been replicated by other practitioners and identified new methodologic questions for practitioners. In some cases, student HIAs have informed government-agency-led HIAs or other health analyses on the same subject (for example, the California Air Resources Board Cap and Trade Regulations HIA). Other core components of the academic practice at Berkley have provided technical assistance to local health agencies that want to conduct HIAs, develop research methods that can be used in HIAs, and mentor graduate-student research and evaluation in the HIA field.

Another example of intensive involvement in HIA in an academic setting is the Health Impact Assessment Project at UCLA. The project began in 2001 with an assessment of the potential avenues for the development of HIA, either

---

[13]See http://www.healthypeople.gov/2020/default.aspx.

as part of or parallel to EIA, with the identification of specific protocols and methods that could be easily and productively adapted from EIA and other fields for use in HIA and with the development of prototype HIAs of policies at federal, state, and local levels (UCLA PH 2011). Since then, the UCLA HIA project has continued to produce demonstration HIAs—HIAs that are produced to demonstrate what an HIA would look like but that are not submitted to decision-makers—across a broad spectrum of policy sectors, including proposed agriculture, education, labor, and planning policies. Collaborating with CDC, the American Planning Association, and Human Impact Partners, the UCLA HIA team has provided HIA training workshops for public agencies and nonprofit organizations. With the aim of lowering the technical barriers to HIA and disseminating HIA practice, they have also developed the HIA Clearinghouse Learning and Information Center (UCLA HIA-CLIC 2011), which includes an archive of completed HIAs in the United States, an extensive explanation of HIA methods, and background literature.

INDIGENOUS PEOPLES

Indigenous peoples in the United States and Canada share a number of factors: they enjoy a close relationship to and a continuing reliance on natural resources for food and subsistence, they have been subject to increased exposure to environmental pollution, they are in the midst of extensive sociocultural change and the strain that it entails, and they experience higher mortality and disease incidence than the general U.S. population (Williams 2010). Those factors suggest that HIA may be an important approach for indigenous peoples. Outside the United States, there have been cases in which non-Western systems of knowledge have been incorporated into HIA, and indigenous peoples and traditional ways of thinking have played an active role in the HIA process. There are both similarities and differences between indigenous peoples' approaches to knowledge and Western impact assessment. The following are some examples of how indigenous peoples around the world have been involved in or used HIA.

- *New Zealand.* In 2005, the Public Health Advisory Committee issued guidance stating that new policies should be appraised for their attention to the principles of the Treaty of Waitangi: partnership, participation, and protection and consequent effects on the health and well-being of Māori Whānau families and communities (PHAC 2005). In 2007, the Ministry of Health published an HIA guide specifically to support Māori health and well-being and to reduce disparities in health (MOH 2007).
- *Australia.* In New South Wales, "health is defined as not just the physical well-being of the individual but the social, emotional and cultural well-being of the whole community" (NSW DH 2007, p. 5). The government requires agencies to submit aboriginal health impact statements with new health-policy proposals for major health strategies and programs and with new health-policy evaluations

(NSW DH 2007). The Australian Indigenous Doctors' Association (AIDA) used HIA to examine critically and refine a sensitive and controversial response from the national government regarding child protection in aboriginal communities in the Northern Territories (AIDA/CHETRE 2010). AIDA/CHETRE (2010) stated that, in addition to drawing on a wide array of expertise and literature, the HIA sought to include the aboriginal and Torres Strait islander peoples' voices, experiences, and knowledge to produce a document meaningful to all stakeholders involved.

• *Thailand.* The WHO definition of health was augmented with the concept of spiritual health (Phoolcharoen et al. 2003).

• *Canada:* The Canadian Handbook on HIA places great premium on aboriginal health and traditional knowledge (Health Canada 2004a,b,c,d).

The People Assessing Their Health (PATH) process can include community HIA and has been used in rural communities in Canada (and in rural and tribal communities in India). It is an inclusive approach that focuses on enabling members of a community to examine a proposal and to present their views to decision-makers. It has been used to examine tourism initiatives and changes in services. One of its main aims is to develop community skills and confidence, and reviews of the process have reported favorable comments from participants (Eaton et al. 2009; Cameron et al. 2011). Eaton et al. (2009) and Cameron et al. (2011) note that it is not always clear how much influence the community HIA has on the decision process. The lack of certainty about effects is shared with other participation methods that are outside the formal decision-making process.

An account of an integrated impact assessment in Alberta shows how understanding and insight of First Nation Peoples was integral to the assessment (Orenstein et al. 2010). Community advisers were hired as part of the integrated impact assessment team. They were able to spend time with the local community, to show respect to and learn from the elders, and to act as a conduit between the external consultants, the elders, and the wider community. The integrated impact assessment results were reported in a summary document that was also published in Cree, the language of the indigenous community. The summary document was based on a question-and-answer format, and it responded to questions that had been raised in consultation with the elders and the wider community.

Despite the examples described and with reference to Canada, Kwiatkowski (2011) states that indigenous communities are rarely engaged in impact assessments undertaken by academe, industry, or government officials. Although tribal environmental policy acts have been enacted by several tribes in the United States and may provide a mechanism for including HIA, it appears that today Alaska is the only state in which American Indian tribes have conducted HIA work. In Alaska, tribal organizations, tribes, and municipal governments have worked within the federal EIS process to pioneer the use of HIA

(Wernham 2007; Bhatia and Wernham 2008). Tribal organizations and the federal agencies leading EISs have worked together to integrate health into the EISs. It is becoming part of accepted practice in Alaska. There are several EISs for which HIAs are planned or in progress, and a working group involving tribal organizations, municipal and state health-department representatives, and federal agencies is developing guidance for HIA in Alaska (Wernham 2009).

PRIVATE SECTOR

Large lending institutions have played a central role in driving HIA in the private sector, but other large corporations are also increasingly adopting standards for HIA in project planning, particularly for natural-resources development. Several multinational oil companies have developed internal corporate standards for HIA or for environmental, social, and health impact assessment (IPIECA/OGP 2007, ICMM 2010; Chevron 2011). Trade associations—including the International Petroleum Industry Environmental Conservation Association and the International Council on Mining and Metals—have also recently developed guides for HIA (IPIECA/OGP 2005; ICMM 2010). The increase in use of HIA by large industries is undoubtedly related to emerging lending standards, which were discussed above. A business case for HIA has also been described and includes following ethical and sustainable development principles, obtaining a social license to operate, ensuring a healthy workforce, reducing conflict in and among local governments and communities, and managing risk (Birley 2005).

Little is known regarding industry standards and practices related to public disclosure and dissemination of HIA results. Committee members have heard of the progress or completion of private-sector HIAs in the United States and internationally, but much of this work is not available through internet searches or on request from the consultants who led the HIAs. The reports appear to remain confidential documents used for planning purposes by a corporation or part of the loan application and verification process. Some corporations may voluntarily make their HIA reports public, but they are generally not required to do so under U.S. law. Consequently, it is difficult to assess the amount of HIA activity or the impact that these HIAs are having on private-sector decisions.

LESSONS LEARNED

Review of the international HIA experience and the current status of HIA in the United States assisted the committee in its task of developing a framework and guidance for HIA in the United States. As a result of its review, the committee made several observations, noted below, that shaped its conclusions and recommendations that are provided in the body of its report.

International Experience

- Legislation that has made HIA a formal requirement has played a key role in advancing HIA practice and making it part of the approval mechanism in many countries. A lack of such requirements has often led to uneven application; as political views have changed, HIA has been discontinued, or resources for conducting it have been reduced as was the case in Canada.
- A legal requirement, however, is not necessarily sufficient for successful implementation of HIA. Examples from Thailand, Québec, and the EU point to the importance of establishing mechanisms for generating knowledge about the health implications of sector policies and for transferring that knowledge to the sectors. Learning about health, its determinants, and policies that can protect health is central to the acceptance and effective use of HIA.
- Standards or minimum requirements for conducting HIA are important for its advancement and inclusion in decision-making. Lack of guidance has sometimes led to minimal health analyses, especially when HIA has been incorporated into EIA, SEA, or other integrated assessment frameworks.
- The international experience demonstrates that having adequate capacity for conducting HIA (expertise and resources) is essential for its success and credibility. The European experience highlights the vital role of capacity-building in the educational system. The EU distributes grants to enable research on HIA methods, and England and Wales have developed courses in public-health departments and universities to help to build the professional foundation for implementing HIA in public and private sectors. Furthermore, centers of excellence and trusted institutions in various countries have played an important role in capacity-building by developing evidence, tools, and guidance that takes into account the business practices of specific sectors.
- Clarity on the allocation of resources for HIA and identification of the entity that will cover the cost is key. In many countries, such as Finland, the responsibility for HIA was passed to communities without the necessary clarity about how to fund it.
- Communication between various fields of expertise has proved important for the successful implementation of the HIA process. For example, Australia's experience demonstrates the importance of having staff in such sectors as fisheries, housing, and transportation work closely with health authorities. Chronically understaffed departments, however, have made such interchanges challenging. The international experience demonstrates that the typical lack of professional interchange between departments that have health expertise and departments that are actively engaged in promulgating policies, programs, and projects is a serious impediment to effective implementation of HIA.
- The needs of native peoples deserve special attention in the context of HIA. The health of those populations has generally been affected in ways that are not recognized by most decision-makers, and their capacity to engage with health professionals is often low. A high percentage of native peoples experi-

ence a subsistence lifestyle and are substantially affected by development that harms the plants and animals on which they depend for daily living. Furthermore, for at least some native peoples, health is defined broadly—"as the social, emotional and cultural well-being of the whole community" (NSWDH 2007, p. 5).

- Although some form of HIA is increasingly prevalent in some parts of the private sector, the lack of transparency in the process makes it impossible to evaluate their professional integrity and credibility.

Development of Health Impact Assessment in the United States

- The increased use of HIA in some local communities and states in the United States indicates that more value is being placed on it. Demand for HIA initially has come from grassroots activities, and growth of the practice in the medium term will depend somewhat on constituent demand.

- Until recently, the analysis of health impacts in the United States has not been consistently considered in federal polices despite the passage of legislation by Congress, an interpretation by the Supreme Court affirming that health impacts are cognizable under NEPA, and executive orders, regulations, and guidance promulgated by the executive branch that call for analysis of health impacts. Although there are many reasons why the analysis of health impacts has not been a major concern, information, education, and experience related to the integration of HIA into NEPA analysis is beginning to increase.

- As discussed in the context of NEPA, the mere promulgation of a requirement to take health impacts into account is not a sufficient basis for the implementation of HIA. As indicated by the international experience, requirements must be specific about when HIA is required and about the standards under which it should be conducted.

- A number of policies and programs, as a matter of law, fall outside NEPA. They range from policies on school nutrition to congressional legislation. Thus, relying on NEPA and EIA laws applicable at state and municipal levels is inadequate to ensure analysis of all important health impacts in all policy sectors.

- There has been no organized U.S. effort to educate those who could benefit from the wider use of HIA about its value, availability, and capabilities. The historical failure to include health analysis uniformly as a part of mandated EIA may obscure the value of HIA. Communication tools to educate diverse groups of potential users of HIA have not been well developed, and the dissemination of basic materials has been primarily opportunistic rather than comprehensive. In addition, a registry that could provide valuable information on groups that have HIA experience or that can provide advice on the costs, timeframes, and sources of specialized expertise has not been created.

REFERENCES

Abrahams, D., A. Pennington, A. Scott-Samuel, C. Doyle, O. Metcalfe, L. den Broeder, F. Haigh, O. Mekel, and R. Fehr. 2004. European Policy Health Impact Assessment: A Guide, University of Liverpool, England; RIVM, Netherlands; Institute of Public Health, Ireland; loegd, Institute of Public Health, NRW Bielefeld, Germany. Prepared for the Health and Consumer Protection Directorate General, European Commission. May 2004 [online]. Available: http://ec.europa.eu/health/ph_projects/2001/monitoring/fp_monitoring_2001_a6_frep_11_en.pdf [accessed May 16, 2011].

AIDA/CHETRE (Australian Indigenous Doctors' Association and Centre for Health Equity Training, Research and Evaluation). 2010. Health Impact Assessment of the Northern Territory Emergency Response [online]. Available: http://www.aida.org.au/viewpublications.aspx?id=3 [accessed May 27, 2011].

Alaska HSS (Alaska Department of Health and Social Services). 2011. Alaska Health Impact Assessment (HIA) Program. Alaska Department of Health and Social Services, Division on Public Health [online]. Available: http://www.epi.alaska.gov/hia/ [accessed July 29, 2011].

ANTHC (Alaska Native Tribal Health Consortium). 2011. Health Impact Assessment Program. Alaska Native Tribal Health Consortium, Anchorage, AK [online]. Available: http://www.anthc.org/chs/ces/hiap/ [accessed Feb. 4, 2011].

Australian Government. 2009. Environment Protection (Impact of Proposals) Act 1974-Order under Section 6, May 29, 1987. F2009B00167 [online]. Available: http://www.comlaw.gov.au/Details/F2009B00167 [accessed Feb. 7, 2011].

Banken, R. 2001. Strategies for Institutionalising HIA. ECHP Health Impact Assessment Discussion Paper No. 1. European Centre for Health Policy, Brussels. World Health Organization [online]. Available: http://www.euro.who.int/__data/assets/pdf_file/0010/101620/E75552.pdf

Banken, R. 2004. HIA of policy in Canada. Pp. 165-175 in Health Impact Assessment: Concepts, Theory, Techniques and Applications, J. Kemm, J. Parry, and S. Palmer, eds. Oxford: Oxford University Press.

BAPE (Bureau d'audiences publiques sur l'environnement--Bureau of Public Hearing on the Environment). 1983. Program of Aerial Spraying Against the Spruce Budworm [in French]. Investigation Report and Public Hearing No. 11. Office of Public Hearings on Environment, Government of Québec, Sainte-Foy, Canada [online]. Available: www.bape.gouv.qc.ca/sections/rapports/publications/bape011.pdf [accessed May 31, 2011].

Barton, H., M. Grant, C. Mitcham, and C. Tsourou. 2009. Healthy urban planning in European cities. Health Promot. Int. 24(suppl. 1):i91-i99.

Berensson, K. 2004. HIA at the local level in Sweden. Pp. 213-222 in Health Impact Assessment: Concepts, Theory, Techniques and Applications, J. Kemm, J. Parry, and S. Palmer, eds. Oxford: Oxford University Press.

Bhatia, R. 2010. A Guide for Health Impact Assessment. California Department of Public Health. October 2010 [online]. Available: http://www.cdph.ca.gov/pubsforms/Guidelines/Documents/HIA%20Guide%20FINAL%2010-19-10.pdf [accessed Apr. 22, 2011].

Bhatia, R., and M. Katz. 2001. Estimation of health benefits accruing from a living wage ordinance. Am. J. Pub. Health 91(9):1398-1402.

Bhatia, R., and A. Wernham. 2008. Integrating human health into environmental impact assessment: An unrealized opportunity for environmental health and justice. Environ. Health Perspect. 116(8):991-1000.

Bhatia, R., L. Farhang, J. Heller, K. Capozza, J. Melendez, K. Gilhuly, and N. Firestein. 2008. A Health Impact Assessment of the California Health Families, Healthy Workplaces Act of 2008. Oakland, CA: Human Impact Partners and San Francisco Department of Public Health. July 30, 2008 [online]. Available: http://www.humanimpact.org/component/jdownloads/finish/5/72 [accessed Feb. 4, 2011].

Birley, M.H. 1991. Guidelines for Forecasting the Vector-Borne Disease Implications of Water Resources Development. PEEM Guidelines Series 2. WHO/CWS/91.3. Geneva: World Health Organization [online]. Available: http://www.who.int/docstore/water_sanitation_health/Documents/PEEM2/english/peem2toc.htm [accessed July 12, 2011]

Birley, M. 2005. Health impact assessment in multinationals: A case study of the Royal Dutch/Shell Group. Environ. Imp. Assess. Rev. 25(7-8):702-713.

Birley, M.H., and G.L. Peralta. 1992. Guidelines for the health impact assessment of development projects. Environmental Paper No 11. Asian Development Bank, Manila.

Birley, M.H., M. Gomes, and A. Davy. 1997. Health aspects of environmental impact assessment. Environmental Assessment Sourcebook Update No 18. The World Bank, Washington, DC [online]. Available: http://siteresources.worldbank.org/INTSAFEPOL/1142947-1118039106036/20526309/Update18HealthAspectsOfEAJuly1997.pdf [accessed July 12, 2011].

BLM (Bureau of Land Management). 2007. Northeast National Petroleum Reserve-Alaska (NPR-A) Draft Supplemental Integrated Activity Plan/Environmental Impact Statement (IAP/EIS). U.S. Department of the Interior, the Bureau of Land Management [online]. Available: http://www.blm.gov/ak/st/en/prog/planning/npra_general/ne_npra/northeast_npr-a_draft.html [accessed Nov. 30, 2010].

Bond, A. 2004. Lessons from EIA. Pp. 131-142 in Health Impact Assessment: Concepts, Theory, Techniques and Applications, J. Kemm, J. Parry, and S. Palmer, eds. Oxford: Oxford University Press.

Bowen, C. 2004. HIA and policy development in London: Using HIA as a tool to integrate health considerations into strategy. Pp. 235-242 in Health Impact Assessment, J. Kemm, J. Parry, and S. Palmer, eds. Oxford, England: Oxford University Press.

Breeze, C., and J. Kemm. 2000. The Health Potential of the Objective 1 Programme for West Wales and the Valleys: A Preliminary Health Impact Assessment. Health Promotion Division, The National Assembly for Wales, Wales [online]. Available: http://www.wales.nhs.uk/sites3/Documents/522/Objective1_full_hia.pdf [accessed May 31, 2011].

British Medical Association. 1998. Health and Environmental Impact Assessment. London: Earthscan.

Cagle, C. 2010. MassDOT Priority: Health Transportation. January 25, 2010 [online]. Available: http://transportation.blog.state.ma.us/blog/2010/01/massdot-priority-healthy-transportation.html [accessed Feb. 4, 2011].

Cameron, C., S. Ghosh, and S.L. Eaton. 2011. Facilitating communities in designing and using their own community health impact assessment tool. Environ. Impact Assess. Rev. 31(4):433-437.

Cave, B., A. Bond, and A. Coutts. 2007. Addressing health in strategic environmental assessment. Town and Country Planning 76(2):59-61.

CEC (Commission of the European Communities). 2007. Together for Health: A Strategic Approach for the EU 2008-2013, White Paper. COM(2007) 630 final. Commission of the European Communities, Brussels, October 23, 2007 [online]. Available: http://ec.europa.eu/health/ph_overview/Documents/strategy_wp_en.pdf [accessed May 31, 2011].

CEQ (Council on Environmental Quality). 1997. Environmental Justice: Guidance under the National Environmental Health Policy Act. Council on Environmental Quality, Executive Office of the President, Washington, DC. December 10, 1997 [online]. Available: http://ceq.hss.doe.gov/nepa/regs/ej/justice.pdf [accessed May 20, 2011].

Chevron. 2011. Stakeholder Engagement. Growing Successful Partnerships. Highlights [online]. Available: http://www.chevron.com/globalissues/corporateresponsibility/2007/stakeholderengagement/#b2 [accessed Feb. 10, 2011].

Corburn, J. 2009. Toward the Healthy City: People, Places, and the Politics of Urban Planning. Cambridge, MA: MIT Press.

Corburn, J., and R. Bhatia. 2007. Health impact assessment in San Francisco: Incorporating the social determinants of health into environmental planning. J. Environ. Plan. Manage. 50(3):323-341.

Cole, B.L., S. Hoffman, and R. Shimkhada. 2007. Health Impact Assessment of Modifications to the Trenton Farmers'Market (Trenton, New Jersey). UCLA Health Impact Group, University of California Los Angelos School of Public Health. March 20, 2007 [online]. Available: http://www.ph.ucla.edu/hs/health-impact/docs/FarmersMktFullReport.pdf [accessed Feb. 11, 2011].

Colson, E. 1971. Social Consequences of Resettlement: The Impact of the Kariba Resettlement Upon the Gwembo Tonga. Manchester, UK: Manchester University Press.

Council of the European Union. 1985. Council Directive of 27 June 1985 on the assessment of the effects of certain public and private projects on the environment. 85/337/EEC. O.J. Eur. Comm. L 175:40-48 [online]. Available: http://ec.europa.eu/environment/eia/full-legal-text/85337.htm [accessed May 31, 2011].

COWI. 2009. Study Concerning the Report on the Application and Effectiveness of the EIA Directive, Final report. European Commission DG ENV. June 2009 [online]. Available: http://ec.europa.eu/environment/eia/pdf/eia_study_june_09.pdf [accessed May 31, 2011].

CSDH (Commission on Social Determinants of Health). 2008. Closing the Gap in a Generation: Health Equity through Action on the Social Determinants of Health. Geneva: World Health Organization [online]. Available: http://www.who.int/social_determinants/thecommission/finalreport/en/index.html [accessed May 11, 2011].

Dannenberg, A.L., R. Bhatia, B.L. Cole, S.K. Heaton, J.D. Feldman, and C.D. Rutt. 2008. Use of health impact assessment in the U.S.: 27 case studies, 1999-2007. Am. J. Prev. Med. 34(3):241-256.

Davies, K., and B. Sadler. 1997. Environmental Assessment and Human Health: Perspectives, Approaches, and Future Directions. Ottawa: Health Canada [online]. Available: http://dsp-psd.pwgsc.gc.ca/Collection/H46-3-7-1997E.pdf [accessed May 20, 2011].

Diwan, V., M. Douglas, I. Karlberg, J. Lehto, G. Magnusson, and A. Ritsatakis, eds. 2000. Health Impact Assessment: From Theory to Practice. Report on the Leo Kaprio Workshop, October 28–30, 1999, Göteborg. NHV-Report 2000:9. WHO Europe, Nordic School of Public Health.

Dora, C. 2004. HIA in SEA and its application to policy in Europe. Pp. 403-410 in Health Impact Assessment: Concepts, Theory, Techniques and Applications, J. Kemm, J. Parry, and S. Palmer, eds. Oxford: Oxford University Press.

Dora, C., and F. Racioppi. 2003. Including health in transport policy agendas: The role of health impact assessment analyses and procedures in the European experience. Bull. World Health Organ. 81(6):399-403.

Eaton, S., L. St-Pierre, and M.C. Ross. 2009. Influencing Healthy Public Policy with Community Health Impact Assessment, National Collaborating Centre for Healthy Public Policy, Canada. November 2009 [online]. Available: http://www.ncchpp.ca/docs/PATH_Rapport_EN.pdf [accessed June 1, 2011].

EC (European Communities). 1992. Treaty on European Union signed at Maastricht on February 7, 1992. O.J. Eur. Comm. C 191, July 29, 1992 [online]. Available: http://eur-lex.europa.eu/en/treaties/dat/11992M/htm/11992M.html#0001000001 [accessed June 1, 2011].

EC (European Communities). 1997. Treaty of Amsterdam amending the Treaty on European Union, the Treaties Establishing the European Communities and related acts. O.J. Eur. Comm. C 340, November 10, 1997 [online]. Available: http://eur-lex.|europa.eu/en/treaties/dat/11997M/htm/11997M.html#0145010077 [accessed June 1, 2011].

EC (European Communities). 2007. Treaty of Lisbon amending the Treaty on European Union and the Treaty establishing the European Community. O.J. EU C 306(50), December 17, 2007 [online]. Available:

EC (European Commission). 2009a. Commission Impact Assessment Guidelines [online]. Available: http://ec.europa.eu/governance/impact/commission_guidelines/commission_guidelines_en.htm [accessed June 1, 2011].

EC (European Commission). 2009b. The Main Changes in the 2009 Impact Assessment Guidelines Compared to 2005 Guidelines. Memo from Secretariat General Unit C/2, Better Regulation and Impact Assessment, September 2, 2009 [online]. Available at http://ec.europa.eu/governance/impact/commission_guidelines/docs/revised_ia_guidelines_memo_en.pdf [accessed June 1, 2011].

EC (European Commission). 2011a. Departments (Directorates-General) and Services. Europa [online]. Available: http://ec.europa.eu/about/ds_en.htm [accessed June 1, 2011].

EC (European Commission). 2011b. Impact Assessment. Europa [online]. Available: http://ec.europa.eu/governance/impact/index_en.htm [accessed June 1, 2011].

EC (European Commission). 2011c. Environmental Assessment. Europa [online]. Available: http://ec.europa.eu/environment/eia/home.htm [accessed June 1, 2011].

EC (European Commission). 2011d. Strategic Environmental Assessment - SEA. Europa [online]. Available: http://ec.europa.eu/environment/eia/sea-legalcontext.htm [accessed June 1, 2011].

Elinder, L.S., L. Joossens, M. Raw, S. Andreasson, and T. Lang. 2003. Public Health Aspects of the EU Common Agricultural Policy. Developments and Recommendations for Change in Four Sectors: Fruit and Vegetables, Dairy, Wine and Tobacco. Swedish National Institute of Public Health [online]. Available: http://www.fhi.se/PageFiles/4464/eu_inlaga.pdf [accessed Feb. 11, 2011].

enHealth (enHealth Council). 2001. Health Impact Assessment Guidelines. Canberra: Commonwealth of Australia. September 2001 [online]. Available: http://www.health.gov.au/internet/main/publishing.nsf/content/35F0DC2C1791C3A2CA256F1900042D1F/$File/env_impact.pdf [accessed May 5, 2011].

EP/Council (European Parliament and Council of the European Union). 2001. Directive 2001/42/EC of the European Parliament and of the Council of 27 June 2001 on the assessment of the effects of certain plans and programmes on the environment. O.J. Eur. Comm. L 197:30-37.

EP/Council (European Parliament and the Council of the European Union). 2003. Directive 85/337/EEC on the assessment of the effects of certain public and private projects on the environment, as amended by Directive 97/11/EC and Directive 2003/35/EC. O.J. EU. L 156:17. June 25, 2003.

EP/Council (European Parliament and Council of the European Union). 2009. Directive 2009/31/EC of the European Parliament and of the Council of 23 April 2009 on the geological storage of carbon dioxide and amending Council Directive 85/337/EEC, European Parliament and Council Directives 2000/60/EC, 2001/80/EC, 2004/35/EC, 2006/12/EC, 2008/1/EC and Regulation (EC) No 1013/2006. O.J. EU L140/141, June 5, 2009 [online]. Available: http://eur-lex.europa.eu/LexUriServ/LexUriServ.do?uri=OJ:L:2009:140:0114:0135:EN:PDF [accessed June 1, 2011].

EPA (U.S. Environmental Protection Agency). 2009. Red Dog Mine Extension Aqqaluk Project. Final Supplemental Environmental Impact Statement. Prepared for U.S. Environmental Protection Agency, Seattle, WA, by Tetra Tech, Inc., Anchorage, AK. October 2009 [online]. Available: http://www.reddogseis.com/Docs/Final/Front_Matter.pdf [accessed Nov. 30, 2010].

EPA (U.S. Environmental Protection Agency). 2010. Scoping a Health Impact Assessment (HIA for the Ports of Los Angeles and Long Beach. Pacific Southwest Region 9, U.S. Environmental Protection Agency [online]. Available: http://www.epa.gov/region9/nepa/PortsHIA/index.html [accessed Nov. 30, 2010].

EU (European Union). 2009. Building Health Communities. Newsletter No 3, November 2009 [online]. Available: http://urbact.eu/fileadmin/Projects/Building_Healthy_Communities_BHC_/outputs_media/newsletter03_1.pdf [accessed Nov. 30, 2010].

EU (European Union). 2011a. About EU: Countries. Europa [online]. Available: http://europa.eu/about-eu/countries/index_en.htm [accessed June 1, 2011].

EU (European Union). 2011b. The 27 Member Countries of the European Union. Europa [online]. Available: http://europa.eu/about-eu/27-member-countries/index_en.htm [accessed June 1, 2011].

Farhang, L., R. Bhatia, C.C. Scully, J. Corburn, M. Gaydos, and S. Malekafzali. 2008. Creating tools for healthy development: Case study of San Francisco's Eastern Neighborhoods Community Health Impact Assessment. J. Pub. Health Manag. Pract. 14(3):255-265.

Farhang, L., and R. Bhatia. 2007. Eastern Neighborhoods Community Health Impact Assessment, Final Report. San Francisco Department of Public Health. September 2007 [online]. Available: http://www.sfphes.org/ENCHIA.htm [accessed Feb. 4, 2011].

Fehr, R., O. Mekel, and R. Welteke. 2004. HIA: The German perspective. Pp. 253-264 in Health Impact Assessment: Concepts, Theory, Techniques and Applications, J. Kemm, J. Parry, and S. Palmer, eds. Oxford: Oxford University Press.

Fischer, T.B. 2010. The consideration of health in SEA Pp. 20-42 in Health and Strategic Environmental Assessment. Background Information and Report of the WHO Consultation Meeting, June 8-9, 2009, Rome, Italy, J. Nowacki, M. Martuzzi, and T.B. Fischer, eds. Denmark: World Health Organization.

Gagnon, F., M. Michaud, S. Termblay and V. Turcotte. 2008. Health Impact Assessment and Public Policy Formulation. Québec Library of National Archives, Québec, Canada [online]. Available: http://www.gepps.enap.ca/GEPPS/docs/EnglishPublications/eis_vf20fev09ang2.pdf [accessed June 1, 2011].

Government of Québec. 1998. The Politics of Health and Well-Being [in French]. Government of Québec, Canada [online]. Available: http://publications.msss.gouv.qc.ca/acrobat/f/documentation/1992/92_713.pdf [accessed Mar. 3, 2011].

Government of Québec. 1999. National Public Health Priorities 1997-2002. Towards Achieving the Expected Results: 1st Report [in French]. Government of Québec, Canada [online]. Available: http://publications.msss.gouv.qc.ca/acrobat/f/documen tation/1998/98-201.pdf [accessed Mar. 3, 2011].

Government of Québec. 2001. Public Health Act. Chapter S-2.2: Article 54. Government of Québec [online]. Available: http://www2.publicationsduquebec.gouv.qc.ca/dyna micSearch/telecharge.php?type=2&file=/S_2_2/S2_2_A.html [accessed June 2, 2011].

Government of Tasmania. 1994. Environmental Management and Pollution Control Act (EMPCA). Government of Tasmania [online]. Available: http://www.environment. tas.gov.au/index.aspx?base=365 [accessed June 8, 2011].

Gulis, G., P. Gry, M.W. Fredsgaard, J. Hindhede, H. Charsmar, A. Wagner Pedersen, A. Dokkedal, and M. Martuzzi. 2008. Health Impact Assessment in New Member States Accession and Pre-Accession Countries (HIA-NMAC), Final Technical Report. University of Southern Denmark, Esbjerg, Denmark [online]. Available: http://ec.europa.eu/health/ph_projects/2004/action1/docs/action1_2004_frep_20_e n.pdf [accessed June 2, 2011].

Harris, P., and J. Spickett. 2011. Health impact assessment in Australia: A review and directions for progress. Environ. Impact Assess. Rev. 31(4):425-432.

Hassan, A.A., M.H. Birley, E. Giroult, R. Zghondi, M.Z. Ali Khan, and R. Bos. 2005. Environmental Health Impact Assessment of Development Projects: A Practical Guide for the WHO Eastern Mediterranean Region. Geneva: World Health Organization [online]. Available: http://www.who.int/water_sanitation_health/resour ces/emroehiabook.pdf [accessed July 12, 2011].

Health Canada. 2004a. Canadian Handbook on Health Impact Assessment, Vol. 1. The Basics. Health Canada. November 2004 [online]. Available: http://dsp-psd.pwgsc. gc.ca/Collection/H46-2-04-343E.pdf [accessed June 2, 2011].

Health Canada. 2004b. Canadian Handbook on Health Impact Assessment, Vol. 2. Approaches and Decision-Making. Health Canada. November 2004 [online]. Available: http://www.hiaconnect.edu.au/files/hia-Volume_2.pdf [accessed June 2, 2011].

Health Canada. 2004c. Canadian Handbook on Health Impact Assessment, Vol. 3. The Multidisciplinary Team. Health Canada. November 2004 [online]. Available: http://dsp-psd.pwgsc.gc.ca/Collection/H46-2-04-362E.pdf [accessed June 2, 2011].

Health Canada. 2004d. Canadian Handbook on Health Impact Assessment, Vol. 4. Health Impacts by Industry Sector. Health Canada. November 2004 [online]. Available: https://files.pbworks.com/download/CPg0409alD/healthimpactassessment/173240 22/Canadian%20Handbook%20of%20HIA%20Vol%204%20health%20impacts% 20by%20industry%20sector%20-%20HC%20Canada%20-%202004.pdf [accessed June 2, 2011].

HEIMTSA Consortium. 2010. Health and Environment Integrated Methodology and Toolbox for Scenario Assessment (HEIMTSA) [online]. Available: http://www.hei mtsa.eu/ [accessed April 4, 2010].

Heller, J., J. Lucky, and W.K. Cook. 2009. Crossings at 29th St./San Pedro St. Area Health Impact Assessment. Human Impact Partners, Oakland, California [online]. Available: http://www.hiaguide.org/sites/default/files/Crossings_LA_29thSt_HIA_ FullReport.pdf [accessed May 17, 2011].

Héroux de Sève, J., D. Talbot, C. Druet, and M. Pigeon. 2008. Review and Outlook 2002-2007-At the Frontier of the Responsibilities of Departments: The Application of Article 54 of the Act on Public Health [in French]. Government of Québec,

Canada [online]. Available: http://publications.msss.gouv.qc.ca/acrobat/f/documentation/2008/08-245-02.pdf [accessed Mar. 3, 2011].

Hilding-Rydevik, T., S. Vohra, A. Ruotsalainen, A. Pettersson, N. Pearce, C. Breeze, M. Hrncarova, Z. Liekovska, K. Paluchova, L. Thomas, and J. Kemm. 2005. Health Aspects in EIA. European Commission Sixth Framework Program [online]. Available: http://www.umweltbundesamt.at/fileadmin/site/umweltthemen/UVP_SUP_EMAS/IMP/IMP3-Health_Aspects_in_EIA.pdf [accessed May 23, 2011].

HM Government of Great Britain. 2007. Greater London Authority Act [online]. Available: www.opsi.gov.uk/Acts/acts2007/pdf/ukpga_20070024_en.pdf [accessed June 2, 2011].

Human Impact Partners. 2009. Concord Naval Weapons Station Reuse Project Health Impact Assessment. Human Impact Partners. January 2009 [online]. Available: http://hiaguide.org/hia/concord-naval-weapons-station-reuse-project-health-impact-assessment [accessed Feb. 4, 2011].

ICMM (International Council on Mining and Metals). 2010. Good Practice Guidance on Health Impact Assessment. International Council on Mining and Minerals [online]. Available: http://www.icmm.com/library/hia [accessed July 12, 2011].

IFC (International Finance Corporation). 2006. Community Health, Safety and Security. Performance Standard 4. International Finance Corporation, April 30, 2006 [online]. Available: http://www.ifc.org/ifcext/sustainability.nsf/AttachmentsByTitle/pol_PerformanceStandards2006_PS4/$FILE/PS_4_CommHealthSafetySecurity.pdf [accessed July 12, 2011]

IFC (International Finance Corporation). 2007. General EHS Guidelines: Community Health and Safety. Environmental, Health, and Safety (EHS) Guidelines. International Finance Corporation, April 30, 2007 [online]. Available: http://www.ifc.org/ifcext/sustainability.nsf/AttachmentsByTitle/gui_EHSGuidelines2007_General EHS_3/$FILE/3+Community+Health+and+Safety.pdf [accessed July 12, 2011].

IFC (International Finance Corporation). 2009. Introduction to Health Impact Assessment. International Finance Corporation, April 2009 [online]. Available: http://www.ifc.org/ifcext/sustainability.nsf/Content/Publications_Handbook_HealthImpactAssessment [accessed July 12, 2011].

INTARESE Consortium. 2010. Integrated Assessment of Health Risks of Environmental Stressors in Europe [online]. Available: http://www.intarese.org/ [accessed April 29, 2010].

IPIECA/OGP (International Petroleum Industry Environmental Conservation Association and International Association of Oil and Gas Producers). 2005. A Guide to Health Impact Assessments in the Oil and Gas Industry. International Petroleum Industry Environmental Conservation Association, and International Association of Oil and Gas Producers [online]. Available: http://www.hiaconnect.edu.au/files/HIA_in_OG.pdf [accessed May 17, 2011].

IPIECA/OGP (International Petroleum Industry Environmental Conservation Association and International Association of Oil and Gas Producers). 2007. Health Performance Indicators: A Guide for the Oil and Gas Industry. OGP Report No. 393. International Petroleum Industry Environmental Conservation Association, and International Association of Oil and Gas Producers [online]. Available: http://www.ipieca.org/system/files/publications/HPI.pdf [accessed June 2, 2011].

ISAIAH. 2011. Health Corridor for All. ISAIAH, Minneapolis, MN [online]. Available: http://isaiah-mn.org/Issues/HealthyCorridorforAll.htm [accessed Feb. 4, 2011].

Ison, E. 2009. The introduction of health impact assessment in the WHO European Healthy Cities Network. Health Promot. Int. 24(suppl. 1):i64-i71.

Knutsson, I., and A. Linell. 2010. Review article: Health impact assessment developments in Sweden. Scand. J. Public Health 38(2):115-120.

Kørnøv, L. 2009. Strategic Environmental Assessment as catalyst of healthier spatial planning: The Danish guidance and practice. Environ. Impact Assess. Rev. 29(1): 60-65.

Kwiatkowski, R.E. 2004. Impact assessment in Canada: An evolutionary process. Pp. 309-316 in Health Impact Assessment: Concepts, Theory, Techniques and Applications, J. Kemm, J. Parry, and S. Palmer, eds. Oxford: Oxford University Press.

Kwiatkowski, R.E. 2011. Indigenous community based participatory research and health impact assessment: A Canadian example. Environ. Impact Assess. Rev. 31(4):445-450.

Kwiatkowski, R.E., and M. Ooi. 2003. Integrated environmental impact assessment: A Canadian example. Bull. World Health Organ. 81(6):434-438.

Lalonde, M. 1974. A New Perspective on the Health of Canadians: A Working Document. Ministry of Supply and Services, Government of Canada, Ottawa [online]. Available: http://www.phac-aspc.gc.ca/ph-sp/pdf/perspect-eng.pdf [accessed May 31, 2011].

Laframboise, H.L. 1973. Health policy: Breaking the problem down into more manageable segments. Can. Med Assoc. J. 108(3):388-391.

Lock, K., M. Gabrijelcic-Blenkus, M. Martuzzi, P. Otorepec, P. Wallace, C. Dora, A. Robertson, and J.M. Zakotnic. 2003. Health impact assessment of agriculture and food policies: Lessons learnt from the Republic of Slovenia. Bull. World Health Organ. 81(6):391-398.

London Health Commission. 2011. HIA Publications. London Health Commission [online]. Available: http://www.london.gov.uk/lhc/publications/hia/ [accessed Mar. 8, 2011].

Los Angeles Country Metropolitan Transportation Authority. 2010. I-710 Corridor Project EIR/EIS News Newsletter. January 2010 [online]. Available: http://www.metro.net/projects_studies/I710/images/I-710-Project-Update-Winter-Spring-2010.pdf [accessed Nov. 30, 2010].

Mahoney, M. 2009. Imperatives for Policy Health Impact Assessment: Perspectives, Positions, Power Relations. Ph.D. Thesis, Deakin University, Victoria, Australia.

Mahoney, M., and G. Durham. 2002. Health Impact Assessment: A Tool for Policy Development in Australia. Deakin University, Geelong, Victoria, Australia [online]. Available: http://www.deakin.edu.au/hmnbs/hia/publications/HIA_Final_Report_2003.pdf [accessed June 3, 2011].

Mahoney, M., S. Simpson, E. Harris, R. Aldrich, and J. Stewart-Williams. 2004. Equity Focused Health Impact Assessment Framework. The Australasian Collaboration for Health Equity Impact Assessment (ACHEIA). August 2004 [online]. Available: http://www.hiaconnect.edu.au/files/EFHIA_Framework.pdf [accessed May 17, 2011].

Mall, A., S. Buccino, and J. Nichols. 2007. Drilling Down: Protecting Western Communities from the Health and Environmental Effects of Oil and Gas Production. New York: Natural Resources Defense Council. October 2007 [online]. Available: http://www.nrdc.org/land/use/down/down.pdf [accessed June 3, 2011].

McKay, L. 2000. Making the Lalonde Report: Towards a New Perspective on Health Project. Canadian Policy Research Networks [online]. Available: http://cprn.org/documents/18406_en.pdf [accessed: Apr. 28, 2011].

Mercier, J.R. 2003. Health impact assessment in international development assistance: The World Bank experience. Bull. World Health Organ. 81(6):461-462.

Milio, N. 1981. Promoting Health Through Public Policy. Philadelphia, PA: F.A. Davis Company.

MMS (Minerals Management Service). 2007a. Outer Continental Shelf Oil and Gas Leasing Program: 2007-2010. Final Environmental Impact Statement, Vol. 1. OCS EIS/EA MMS2007-003. U.S. Department of the Interior, Minerals Management Service, Herndon, VA. April 2007 [online]. Available: http://www.boemre.gov/5-year/2007-2012FEIS/Intro.pdf [accessed Nov. 30, 2010].

MMS (Minerals Management Service). 2007b. Chukchi Sea Planning Area Oil and Gas Sale 193 and Seismic Surveying Activities in the Chukchi Sea. Final Environmental Impact Statement. OCS EIS/EA MMS2007-026. U.S. Department of the Interior, Minerals Management Service, Alaska OCS Region [online]. Available: http://alaska.boemre.gov/ref/EIS%20EA/Chukchi_FEIS_193/feis_193.htm [accessed Nov. 30, 2010].

MOH (Ministry of Health). 2007. Whanau Ora Health Impact Assessment. Ministry of Health [online]. Available: http://www.moh.govt.nz/moh.nsf/pagesmh/6022/$File/whanau-ora-hia-2007.pdf [accessed June 27, 2011].

NCCHPP (National Collaborating Centre for Healthy Public Policy). 2008. The Québec Public Health Act's Section 54: Briefing Note. Preliminary version - for discussion. National Collaborating Centre for Healthy Public Policy, Québec. March 2008 [online]. Available http://www.ncchpp.ca/docs/Section54English042008.pdf [accessed June 3, 2011].

NHC Thailand (National Health Commission Office of Thailand). 2007. National Health Act, B.E. 2550 (A.D. 2007). National Health Commission Office of Thailand [online]. Available: http://en.nationalhealth.or.th/sites/default/files/fromNHCThailand/data/HealthAct07.pdf [accessed July 6, 2011].

NHC Thailand (National Health Commission Office of Thailand). 2008. HIA for HPP towards Healthy Nation: Thailand's Recent Experiences. National Health Commission Office of Thailand [online]. Available: http://en.nationalhealth.or.th/sites/default/files/fromNHCThailand/data/HIA%20for%20HPP_Final%20Version_pdf [accessed July 6, 2011].

NHMRC (National Health and Medical Research Council). 1994. Framework for Environmental and Health Impact Assessment. National Health and Medical Research Council, Commonwealth of Australia, Canberra [online]. Available: http://www.hiaconnect.edu.au/files/NHMRC_EHIA_Framework.pdf [accessed June 3, 2011].

Nilunger, L., L.S. Elinder, and B. Pettersson. 2003. Health Impact Assessment: Screening of Swedish governmental inquiries. Eurohealth 8(5):30-32.

Noble, B.F., and J.E. Bronson. 2005. Integrating human health into environmental impact assessment: Case studies of Canada's Northern Mining Resource Sector. Arctic 58(4):395-405.

Noble, B.F., and J.E. Bronson. 2006. Practitioner survey of the state of health integration in environmental assessment: The case of northern Canada. Environ. Impact Assess. Rev. 26(4):410-424.

NPHP (National Public Health Partnership). 2005. Health Impact Assessment: Legislative and Administrative Frameworks. National Public Health Partnership, Melbourne, Australia [online]. Available: http://www.dhs.vic.gov.au/nphp/workprog/lrn/legtools/hia_legframe.pdf [accessed June 3, 2011].

NPHPPHC (National Prevention, Health Promotion and Public Health Council). 2010. The National Prevention and Health Promotion Strategy, Draft Framework. National Prevention, Health Promotion and Public Health Council. October 1, 2010 [online]. Available: http://www.healthcare.gov/center/councils/nphpphc/draftframework_pdf [accessed Nov. 30, 2010].

NSW DH (New South Wales Department of Health). 2007. Aboriginal Health Impact Statement and Guidelines. PD2007-082. Department of Health, New South Wales Government, Australia. November 12, 2007 [online]. Available: http://www.health.nsw.gov.au/policies/pd/2007/pdf/PD2007_082.pdf [accessed Nov. 30, 2010].

O'Mullane, M. 2011. Health Impact Assessment (HIA) and Institutionalisation: The Slovak Experience. Presentation at XI HIA International Conference, April 14-15, Granada, Spain.

Opinion Leader Research. 2003. Report on the Qualitative Evaluation of Four Health Impact Assessments on Draft Mayoral Strategies for London, London Health Commission, Greater London Authority and the London Health Observatory. Opinion Leader Research, London. August 2003 [online]. Available: http://www.hiaconnect.edu.au/files/London_Mayoral_HIAs_Evaluation.pdf [accessed May 23, 2011].

Oregon Government. 2011. Health Impact Assessment [online]. Available: http://www.oregon.gov/DHS/ph/hia/index.shtml [accessed Feb. 4, 2011].

Orenstein, M., T. Fossgard-Moser, T. Hindmarch, S. Dowse, J. Kuschminder, P. McCloskey, and R.K. Mugo. 2010. Case study of an integrated assessment: Shell's north field test in Alberta, Canada. IAPA 28(2):147-157.

PHAC (Public Health Advisory Committee). 2005. A Guide to Health Impact Assessment: A Policy Tool for New Zealand, 2nd Ed. Wellington, New Zealand: PHAC [online]. Available: http://www.phac.health.govt.nz/moh.nsf/pagescm/764/$File/guidetohia.pdf [accessed May 9, 2011].

Phoolcharoen, W., D. Sukkumnoed, and P. Kessomboon. 2003. Development of health impact assessment in Thailand: Recent experiences and challenges. Bull. World Health Organ. 81(6):465-467.

Poirier, A. 2011a. Institutionalizing HIA in Québec: Section 54 of the Public Health Act. Presentation at XI HIA International Conference April 14-15, Granada, Spain [online]. Available: http://si.easp.es/eis2011/wp-content/uploads/2011/04/English-presentation-GrenadaInstitutionnalisationEIS-1.ppt [accessed June 3, 2011].

Poirier, A. 2011b. Beyond Health Impact Assessment: A Government Policy for Health and Well-Being. Presentation at XI HIA International Conference April 14-15, Granada, Spain [online]. Available: http://si.easp.es/eis2011/wp-content/uploads/2011/04/English-presentation_HIA-conference-Grenade-2.pdf [accessed June 3, 2011].

Public and Environmental Health Service. 1998. Manual for Local Government. Department of Health and Human Services, Tasmania, Australia. December 1998.

Ross, C.L. 2007. Atlanta Beltline: Health Impact Assessment. Center for Quality Growth and Regional Development, Georgia Institite of Technology, Atlanta, GA [online]. Available: http://www.healthimpactproject.org/resources/document/Atlanta-Beltline.pdf [accessed May 18, 2011].

SFDPH (San Francisco Department of Public Health). 2011. Assessing the Health Impacts of Road Pricing Policy Proposals. Program on Health Equity and Sustainability, San Francisco Department of Public Health [online]. Available: http://www.sfphes.org/HIA_Tools/RoadPricing_Health_Pathways.pdf [accessed May 24, 2011].

SFCC (Federation of Swedish County Councils). 1998. Focusing on Health: How Can the Health Impact of Policy Decisions be Assessed? Federation of Swedish County Councils, Stockholm, Sweden [online]. Available: http://www.lf.se/hkb/engelskversion/general.htm.

Simpson, S., M. Mahoney, E. Harris, R. Aldrich, and J. Stewart-Williams. 2005. Equity-focused health impact assessment: A tool to assist policy makers in addressing health inequalities. Environ. Impact Assess. Rev. 25(7-8):772-782.

Smith, K.E., G. Fooks, J. Collin, H. Weishaar, and A.B. Gilmore. 2010a. Is the increasing policy use of Impact Assessment in Europe likely to undermine efforts to achieve healthy public policy? J. Epidemiol. Community Health 64(6):478-487.

Smith, K.E., G. Fooks, J. Collin, H. Weishaar, S. Mandal, and A.B. Gilmore. 2010b. "Working the system"- British American tobacco's influence on the European Union treaty and its implications for policy: An analysis of internal tobacco industry documents. PLoS Med. 7(1):e1000202.

Ståhl, T.P. 2010. Is health recognized in the EU's policy process? An analysis of the European Commission's impact assessments. Eur. J. Public Health 20(2):176-181.

Ståhl, T., M. Wismar, W. Ollila, E. Lahtinen, and K. Leppo, eds. 2006. Health in All Policies. Prospects and Potentials. Ministry of Social Affairs and Health, Finland [online]. Available: http://www.euro.who.int/__data/assets/pdf_file/0003/109146/E89260.pdf [accessed June 3, 2011].

Thai Laws. 2007. Constitution of the Kingdom of Thailand B.E. 2550 (2007). Bangkok: Bureau of Printing Service [online]. Available: http://www.senate.go.th/t_senate/English/constitution2007.pdf [accessed Feb. 7, 2010].

Thornton, R.L.J., C.M. Fichtenberg, A. Greiner, B. Feingold, J.M. Ellen, J.M. Jennings, M.A. Shea, J. Schilling, R.B. Taylor, D. Bishai, and M. Black. 2010. Zoning for a Healthy Baltimore: A Health Impact Assessment of the Transform Baltimore Comprehensive Zoning Code Rewrite. Baltimore, MD: Johns Hopkins University Center for Child and Community Health Research. August 2010 [online]. Available: http://www.hopkinsbayview.org/bin/a/n/FullReportBW.pdf [accessed Nov. 30, 2010].

Tiffen, M. 1989. Guidelines for the Incorporation of Health Safeguards into Irrigation Projects through Intersectoral Cooperation, with Special Reference to Vector-Borne Diseases. PEEM Guidelines 1. Geneva: World Health Organization [online]. Available: http://www.who.int/water_sanitation_health/resources/peem1/en/index.html [accessed July 8, 2011].

UCB HIA (University of California, Berkeley Health Impact Assessment Group). 2011. Projects and Research [online]. Available: http://sites.google.com/site/ucbhia/projects-and-research [accessed Feb. 4, 2011].

UCLA HIA-CLIC (University of California, Los Angeles-Health Impact Assessment Clearinghouse Learning and Information Center). 2011. Completed HIAs. University of California, Los Angeles [online]. Available: http://www.hiaguide.org/hias [accessed Feb. 3, 2011].

UCLA PH (University of California Los Angeles, School of Public Health). 2004. Summary of the Health Impact Assessment of the 2002 Federal Farm Bill. Health Impact Assessment Project, Partnership for Prevention/UCLA School of Public Health. December 2, 2004 [online]. Available: http://www.ph.ucla.edu/hs/health-impact/docs/FarmBillSummary.pdf [accessed Nov. 30, 2010].

UCLA PH (University of California Los Angeles, School of Public Health). 2011. About HIA. Health Impact Assessment Project. University of California Los Angeles

School of Public Health [online]. Available: http://www.ph.ucla.edu/hs/health-impact/ [accessed Feb. 4, 2011].
UNECE (United Nations Economic Commission for Europe). 1991. Convention on Environmental Impact Assessment in a Transboundary Context –the Espoo (EIA) Convention, February 25, 1991. United Nations Economic Commission for Europe [online]. Available: http://www.unece.org/env/eia/eia.htm [accessed June 6, 2011].
UNECE (United Nations Economic Commission for Europe). 1998. Aarhus Convention: Convention on Access to Information, Public Participation in Decision-Making and Access to Justice in Environmental Matters, June 25, 1998, Aarhus, Denmark [online]. Available http://www.unece.org/env/pp/documents/cep43e.pdf [accessed June 6, 2011].
UNECE (United Nations Economic Commission for Europe). 2003. Protocol on Strategic Environmental Assessment to the Convention on Environmental Impact Assessment in a Transboundary Context, Extraordinary Meeting of the Parties to the ESPOO Convention, May 21, 2003, Kyiv [online]. Available: http://www.unece.org/env/eia/sea_protocol.htm [accessed Nov. 30, 2010].
WA SBOH (Washington State Board of Health). 2007a. Health Impact Review: Building Bridges for Dropout Reductions. February 1, 2007. Washington State Board of Health, Olympia, WA [online]. Available: http://www.sboh.wa.gov/HIR/docs/HIR_2007-01.pdf [accessed Feb. 1, 2011].
WA SBOH (Washington State Board of Health). 2007b. Health Impact Review: Financial Incentives to Attract Excellent Teachers for Hard-to-Staff Schools and Subjects. February 1, 2007. Washington State Board of Health, Olympia, WA [online]. Available: http://www.sboh.wa.gov/HIR/docs/HIR_2007-02.pdf [accessed Feb. 1, 2011].
WA SBOH (Washington State Board of Health). 2008a. Health Impact Review: Substitute House Bill 1675. Providing Certain Public Notices in Languages Other Than English. February 18, 2008. Washington State Board of Health, Olympia, WA [online]. Available: http://www.sboh.wa.gov/HIR/docs/HIR_2008-01.pdf [accessed Feb. 1, 2011].
WA SBOH (Washington State Board of Health). 2008b. Health Impact Review: Establishing the Financial Services Intermediary. February 22, 2008. Washington State Board of Health, Olympia, WA [online]. Available: http://www.sboh.wa.gov/HIR/docs/HIR_2008-02.pdf [accessed Feb. 1, 2011].
WA SBOH (Washington State Board of Health). 2008c. Health Impact Report: Student Discipline Policies on Restraint. February 15, 2008. Washington State Board of Health, Olympia, WA [online]. Available: http://www.sboh.wa.gov/HIR/docs/HIR_2008-03.pdf [accessed Feb. 1, 2011].
WA SBOH (Washington State Board of Health). 2009. Health Impact Review: Proposed Cuts to Health Care and Human Services Programs. March 31, 2009. Washington State Board of Health, Olympia, WA [online]. Available: http://www.sboh.wa.gov/HIR/docs/HIR_2009-01.pdf [accessed February 1, 2011].
WA SBOH (Washington State Board of Health). 2010. Health Impact Review. House Bill 1341: Motivating Students through Incentives to Pursue Postsecondary Education by Eliminating Statewide Assessments as a High School Graduation Requirement. Washington State Board of Health, Olympia, WA. January 8, 2010 [online]. Available: http://www.sboh.wa.gov/HIR/docs/HIR_2009-03.pdf [accessed Feb. 1, 2011].
Weitzenfeld, H. 1996. Manual Básico de Evaluación de Impacto en el Ambiente y la Salud: de Acciones Proyectadas. Centro Panamericano de Ecología Humana y

Salud, Metepec, Estado de Mexico [online]. Available: http://www.bvsde.ops-oms.org/bvsaia/fulltext/basico/031171-01.pdf [accessed July 11, 2011].

Wernham, A. 2007. Inupiat health and proposed Alaskan oil development: Results of the first Integrated Health Impact Assessment/Environmental Impact Statement of proposed oil development on Alaska's North Slope. EcoHealth 4(4):500-513.

Wernham, A. 2009. Building a statewide health impact assessment program: A case study from Alaska. Northwest Public Health 26(1):16-17.

White House Task Force on Childhood Obesity. 2010. Solving the Problem of Childhood Obesity within a Generation, Report to the President. White House Task Force on Childhood Obesity. May 2010 [online]. Available: http://www.letsmove.gov/sites/letsmove.gov/files/TaskForce_on_Childhood_Obesity_May2010_FullReport.pdf [accessed June 6, 2011].

WHO (World Health Organization). 1986. Ottawa Charter for Health Promotion. First International Conference on Health Promotion, November, 21, 1986, Ottawa. WHO/HPR/HEP/95.1. World Health Organization [online]. Available: http://www.who.int/hpr/NPH/docs/ottawa_charter_hp.pdf [accessed June 6, 2011].

WHO (World Health Organization). 1988. The Adelaide Recommendations on Healthy Public Policy. Second International Conference on Health Promotion, April 5-9, 1988, Adelaide, South Australia. WHO/HPR/HEP/95.2. World Health Organization [online]. Available: http://www.who.int/hpr/NPH/docs/adelaide_recommendations.pdf [accessed June 6, 2011].

WHO (World Health Organization). 2000a. Human Health and Dams: The World Health Organization's Submission to the World Commission on Dams (WCD). Geneva: World Health Organization [online]. Available: http://www.who.int/docstore/water_sanitation_health/vector/dams.htm [accessed July 6, 2011].

WHO (World Health Organization). 2000b. Intersectoral Decision-Making Skills in Support of Health Impact Assessment of Development Projects: Final Report on the Development of a Course Addressing Health Opportunities in Water Resources Development 1988- 1998. Geneva: World Health Organization [online]. Available: http://www.who.int/water_sanitation_health/resources/en/HIADPexec.pdf [accessed July 6, 2011].

WHO (World Health Organization). 2005a. Health Impact Assessment Toolkit for Cities. Document 3. Brochure on How Health Impact Assessment Can Support Decision-Making. EUR/05/5048991. Copenhagen, Denmark: World Health Organization for Europe [online]. Available: http://www.euro.who.int/__data/assets/pdf_file/0016/101509/HIA_toolkit_3.pdf [accessed June 6, 2011].

WHO (World Health Organization). 2005b. Health Impact Assessment: Toolkit for Cities. Document 1. Background Document: Concepts, Processes, Methods. Vision to Action. EUR/05/5048991. Copenhagen, Denmark: World Health Organization for Europe [online]. Available: http://www.bvsde.paho.org/bvsacd/cd32/hia1.pdf [accessed June 6, 2011].

WHO (World Health Organization). 2005c. Health Impact Assessment: Toolkit for Cities. Document 5. Introducing Health Impact Assessment in Bologna, Italy: A Case Study. EUR/05/5048991. Copenhagen, Denmark: World Health Organization for Europe [online]. Available: http://www.bvsde.paho.org/bvsacd/cd32/hia5.pdf [accessed June 6, 2011].

WHO (World Health Organization). 2005d. Health Impact Assessment: Toolkit for Cities. Document 4. Introducing Health Impact Assessment in Trnava, Slovakia: A Case Study. EUR/05/5048991. Copenhagen, Denmark: World Health Organization

Regional Office for Europe [online]. Available: http://www.euro.who.int/_data/assets/pdf_file/0010/101512/HIA_toolkit_4.pdf [accessed June 6, 2011].

WHO (World Health Organization). 2008. Protecting Health in Europe from Climate Change. World Health Organization Regional Office for Europe [online]. Available: http://www.euro.who.int/_data/assets/pdf_file/0016/74401/E91865.pdf [accessed June 6, 2011].

WHO (World Health Organization). 2011. WHO European Healthy Cities Network [online] Available: http://www.euro.who.int/en/what-we-do/health-topics/environmental-health/urban-health/activities/healthy-cities/who-european-healthy-cities-network [accessed Mar. 1, 2011].

WI DPH (Wisconsin Department of Public Health). 2010. Wisconsin Health Impact Assessment Online Toolkit. Considering Health in All Policies. Wisconsin Department of Public Health [online]. Available: http://www.dhs.wisconsin.gov/hia/survey/index.htm [accessed Feb. 4, 2011].

Williams, T. 2010. Health Impact Assessment: A Viable Tool for Protecting Tribal Communities? Presentation at the Second Meeting on Health Impact Assessment, May 12, 2010, Washington, DC.

Wismar, M., J. Blau, K. Ernst, and J. Figueras. 2007. The Effectiveness of Health Impact Assessment: Scope and Limitations of Supporting Decision-Making in Europe. Copenhagen: World Health Organization [online]. Available: http://www.euro.who.int/_data/assets/pdf_file/0003/98283/E90794.pdf [accessed May 18, 2011].

World Bank. 2011. Safeguard Policies. World Bank [online]. Available: http://go.worldbank.org/WTA1ODE7T0 [accessed June 6, 2011].

Appendix B

Biographic Information on the Committee on Health Impact Assessment

Richard J. Jackson (*Chair*) is a professor and chair of environmental health sciences at the University of California, Los Angeles. He has worked extensively on the impact of the environment on public health, and over the last decade much of his work has focused on how the built environment affects health. In 2004, he was co-author of *Urban Sprawl and Public Health*. Dr. Jackson is currently working on policy analyses of environmental impacts on health, from chemical body burdens to climate change to urban design. In addition, he is evaluating the effects of farming, education, housing, and transportation policies on health. Dr. Jackson chaired the American Academy of Pediatrics Committee on Environmental Health and recently served on the Board of Directors of the American Institute of Architects. He serves on the editorial boards of the *American Journal of Industrial Medicine*, *Environmental Research*, and *Public Health Reports*. He is a member of the Institute of Medicine Roundtable on Environmental Health Sciences, Research, and Medicine and of the National Research Council Committee on "Sustainable" Products and Services. Dr. Jackson earned his MD from the University of California, San Francisco.

Dinah Bear is an attorney at law in Washington, DC, and previously served for over 25 years on the president's Council on Environmental Quality (CEQ). She joined CEQ as deputy general counsel in 1981, was appointed general counsel in January 1983, and served in that capacity until October 1993. She resumed that position in January 1995 and was with CEQ until her retirement from government at the end of 2007. At CEQ, she was responsible for interpreting the legal requirements of the National Environmental Policy Act (NEPA) and assisted in overseeing the implementation of NEPA throughout the executive branch. Ms. Bear currently serves on the board of Defenders of Wildlife; Humane Borders, a faith-based organization based in Tucson, Arizona; and the Mt. Graham Coalition, and is an adviser to the Center for International Environmental Law. Ms.

Bear earned her J.D. from McGeorge School of Law and has been admitted to practice by the District of Columbia Bar, the State Bar of California, and the U.S. Supreme Court. She has chaired the American Bar Association's Standing Committee on Environmental Law and the District of Columbia Bar Association's Section on Environment and Natural Resources. She has received the award for Distinguished Achievement in Environmental Law and Policy from the American Bar Association.

Rajiv Bhatia is director of occupational and environmental health for the San Francisco Department of Public Health and holds a clinical appointment at the University of California, San Francisco. He is responsible for developing, implementing, and evaluating environmental health policy in San Francisco. Under Dr. Bhatia's leadership, the Department of Public Health has expanded environmental-health practice to ensure safe and adequate housing, to support worker health rights, to enhance connections between regional agriculture and urban consumers, and to integrate public health and urban planning. As part of those initiatives, the department is developing and evaluating tools for health impact assessment (HIA) and has conducted HIA on local land-use and transportation plans and projects, local and state workplace and employment regulations, regional maritime-port development proposals, and California state climate-change mitigation strategies. Dr. Bhatia developed and currently teaches a graduate course on HIA at the University of California, Berkeley and regularly conducts HIA training for peers; federal, state, and local public institutions; and community organizations. He is a co-founder and scientific director of the nonprofit Human Impact Partners, which conducts HIA and HIA training and facilitation for other organizations. Dr. Bhatia was a founding member of the Health and Social Justice Team for the National Association of County and City Health Officials and is a former board member of the Pesticide Action Network and the Asian Pacific Environmental Network. Dr. Bhatia earned an MD from Stanford University and an MPH from the University of California, Berkeley.

Scott B. Cantor is a professor in the Section of Health Services Research in the Department of Biostatistics of The University of Texas MD Anderson Cancer Center. He also holds adjunct-professor positions in The University of Texas Health Science Center in the Houston School of Public Health and Rice University, and he is a faculty member for the Program in Biomathematics and Biostatistics of The University of Texas Graduate School of Biomedical Sciences in Houston. Dr. Cantor's research focuses on the theoretical issues concerning cost-effectiveness analysis and diagnostic testing and on clinical issues in cancer prevention, particularly prostatic-cancer screening and cervical-precancer diagnosis. He is a past president of the Society for Medical Decision Making and is a member of the Decision Analysis Society, the Society for Judgment and Decision Making, and the Institute for Operations Research and the Management Sciences. Dr. Cantor earned a PhD in decision sciences from Harvard University.

Ben Cave is chief executive of Ben Cave Associates. He has specialized in health and social impact assessment for the last 13 years. His work has two broad themes. The first addresses health in statutory assessments. On a policy level, he advises the World Health Organization and the UK Department of Health on requirements and methods for strategic environmental assessment. On a project level, he leads health impact assessments in conjunction with environmental assessments in a wide variety of sectors. The second major theme of his work is to improve the consideration of health issues in the wider planning process and the consideration of environmental issues by health stakeholders. Mr. Cave is associated with several professional organizations and is the chair of the Health Section of the International Association for Impact Assessment and an associate member of the Institute of Environmental Management and Assessment. He earned an MSc in health-promotion sciences from the London School of Hygiene and Tropical Medicine.

Ana V. Diez Roux is a professor of epidemiology and director of the Center for Social Epidemiology and Population Health in the School of Public Health, a research professor in the Survey Research Center in the Institute for Social Research, and director of the Robert Wood Johnson Health and Society Scholars Program at the University of Michigan. Dr. Diez Roux has been an international leader in the investigation of the social determinants of health, the application of multilevel analysis in health research, and the study of neighborhood health effects. Her research includes social epidemiology and health disparities, environmental-health effects, urban health, psychosocial factors in health, cardiovascular-disease epidemiology, social environment-gene interactions, and the use of complex systems approaches in population health. She serves on numerous review and advisory committees, was awarded the Wade Hampton Frost Award for her contributions to public health by the American Public Health Association, and was elected to the Institute of Medicine in 2009. Dr. Diez Roux received an MD from the University of Buenos Aires and a master's degree in public health and a doctorate in health policy from the Johns Hopkins School of Hygiene and Public Health.

Carlos Dora is coordinator of a global program on health impact assessment in the Public Health and the Environment Department of the World Health Organization (WHO). He leads a unit on interventions for healthy environments that is focused on improving the health consequences of policies in different sectors of the economy. Earlier, Dr. Dora had developed a program on the environmental health implications of transport policies and worked on policy frameworks for environmental health, including the Strategic Environmental Assessment Protocol and Environment and Health Performance Reviews, and on risk assessment, including assessment related to the Chernobyl disaster and depleted uranium. He also served as a senior policy analyst at the office of the WHO director general. Dr. Dora earned a PhD from the London School of Hygiene and Tropical Medicine.

Jonathan E. Fielding is the director of the Los Angeles County Department of Public Health and the county health officer and is responsible for all public-health functions, such as surveillance and control of both communicable and noncommunicable diseases and health protection, including emergency preparedness, for the county's 10 million residents. He is also a member of the Los Angeles First 5 Commission, which grants over $100 million per year to improve the health and development of children 0-5 years old. Dr. Fielding chairs the U.S. Community Preventive Services Task Force and was a founding member of the U.S. Clinical Preventive Services Task Force. He also chairs the U.S. Department of Health and Human Services Secretary's Advisory Committee on National Health Promotion and Disease Prevention Objectives for 2020 and was appointed to the California Department of Public Health Advisory Board. Dr. Fielding is a professor in the Schools of Medicine and Public Health at the University of California, Los Angeles (UCLA) and the author of over 175 peer-reviewed publications, editorials, and book chapters on public health, health policy, health economics, emergency preparedness, and evidence-based public-health practice issues. He has been the principal investigator on grants to develop health impact assessment methods and to use them in assessing the health effects of existing or proposed policies in other sectors. He is editor of the *Annual Review of Public Health* and chairman of Partnership for Prevention. He also serves on the board of the American Legacy Foundation and is an elected member of the Institute of Medicine. He formerly served as Massachusetts Commissioner of Public Health and vice president of Johnson & Johnson. Dr. Fielding has received numerous awards, including the Sedgwick Memorial Medal from the American Public Health Association, the Distinguished Alumni Achievement Award from the Harvard School of Public Health, and the UCLA Medal, which is the university's highest honor. He received his MD and MPH from Harvard University and an MBA in finance from the Wharton School of Business.

Joshua Graff Zivin is an associate professor of economics in the Graduate School of International Relations and Pacific Studies of the University of California, San Diego (UCSD). He is also a research associate at the National Bureau of Economic Research and research director for international environmental and health studies at the Institute for Global Conflict and Cooperation. From 2004 to 2005, he served as senior economist for health and the environment for the White House Council of Economic Advisers. Before joining the faculty at UCSD, Dr. Graff Zivin was an associate professor of economics in the Mailman School of Public Health of Columbia University. Dr. Graff Zivin's research spans three fields of economics—health, the environment, and international development—and focuses on how uncertainty and heterogeneity affect both individual and societal decision-making. He is currently engaged in three large projects. The first makes use of primary data collected over the last several years to examine the economic impacts of the AIDS crisis in Africa. The second relies on a unique, matched dataset to understand the role of institutions, social networks, and financial incentives in the production of new scientific knowledge in the life sciences. The third examines behavioral responses to poor air quality and its implications for the economic costs of climate change. Dr. Graff Zivin earned his PhD from University of California, Berkeley.

Jonathan I. Levy is professor of environmental health at Boston University School of Public Health. Dr. Levy's research centers on developing models for quantitative assessment of the environmental and health impacts of air pollution from local to national scales, with a focus on urban environments and variability in exposures and risks. Current research efforts involve developing methods for cumulative risk assessment, addressing chemical and nonchemical stressors in a low-income community, modeling spatial and temporal patterns of air pollution associated with traffic and aircraft, and assessing the influence of indoor environmental interventions on pediatric asthma. Dr. Levy was the recipient of the Walter A. Rosenblith New Investigator Award from the Health Effects Institute in 2005. He is a member of the U.S. Environmental Protection Agency Advisory Council on Clear Air Compliance Analysis and previously served as a member of the National Research Council Committee on Improving Risk Analysis Approaches Used by the U.S. Environmental Protection Agency and Committee on the Effects of Changes in New Source Review Programs for Stationary Sources of Air Pollutants. Dr. Levy earned an ScD in environmental science and risk management from the Harvard School of Public Health.

Julia B. Quint is a research scientist and retired as chief of the Hazard Evaluation System and Information Service in the Occupational Health Branch of the California Department of Public Health. She was involved in identifying and evaluating reproductive toxicants, carcinogens, and other workplace chemical hazards and in developing research projects and other strategies to protect workers, communities, and the environment from the hazards of toxic chemicals. Dr. Quint is a member of the California Environmental Contaminant Biomonitoring Program Scientific Guidance Panel and the California Environmental Protection Agency Green Ribbon Science Panel. She was also a member of the National Research Council Committee on Tetrachloroethylene. Dr. Quint received a PhD in biochemistry from the University of Southern California.

Samina Raja is associate professor of urban and regional planning and adjunct associate professor of health behavior at the University at Buffalo, the State University of New York. Her research focuses on planning and design for healthy communities, sustainable food systems, and the fiscal dimensions of planning. Her research on healthy communities examines the influence of the food and built environments on obesity and physical activity. Her interests in fiscal dimensions of planning pertain to the methods that planners use for measuring the fiscal impacts of land development. Dr. Raja's service to the community and the planning profession is linked to her research interests. She is an active member of the Food Interest Group of the American Planning Association and serves on the Board of Directors of the Community Food Security Coalition. Dr. Raja earned a PhD in urban and regional planning from the University of Wisconsin-Madison.

Amy J. Schulz is associate professor in the Department of Health Behavior and Health Education and associate director of the Center for Research on Ethnicity, Culture, and Health of the University of Michigan School of Public Health and associate research professor in the Institute for Research on Women and Gender. Dr. Schulz has a longstanding commitment and research record focused on the contributions of social factors to racial, ethnic, and socioeconomic disparities in health. Her current research focuses on community-based participatory approaches to understanding social inequalities as they influence health disparities with a particular focus on the health of urban residents. Since 2000, her work has focused on understanding social determinants of obesity and cardiovascular disease in Detroit and evaluating the impacts of interventions to reduce them. She is principal investigator for the Lean & Green in Motown Project, which addresses associations between social and physical environments and risk factors associated with obesity and the Community Approaches to Cardiovascular Health intervention research project to improve cardiovascular health. She previously served as co-principal investigator for the Promoting Healthy Eating in Detroit project. In addition to directing a number of major studies of chronic conditions in multiethnic populations, she is a leader in the field of community-based participatory approaches to research and intervention design. She has been a frequent contributor to the published literature on racial and ethnic disparities in health, on contributions of social factors to health disparities, and on the active engagement of representatives of communities disproportionately affected by health risks in researching and developing interventions to improve health. Dr. Schulz received her PhD in sociology and her MPH in health behavior and health education from the University of Michigan.

Aaron A. Wernham is director of the health impact project at Pew Charitable Trusts. The project involves the creation of a new national center to promote the use of health impact assessment (HIA) and support the growth of the field in the United States. Dr. Wernham is a nationally recognized expert who has led HIA at the state and federal level and conducted HIA training for, collaborated with, and advised numerous health and environmental regulatory agencies on integrating HIA into their programs. Earlier, Dr. Wernham was a senior policy analyst with the Alaska Native Tribal Health Consortium, where he led the first successful efforts in the United States to integrate HIA formally into the federal environmental impact statement process. He also directed a collaborative state-tribal-federal working group on HIA and, with the assistance of this group, wrote HIA guidance for federal and state environmental regulatory and permitting efforts. Dr. Wernham received his MD from the University of California, San Francisco.

Appendix C

Statement of Task of the Committee on Health Impact Assessment

An NRC/IOM committee will develop a framework, terminology, and guidance for conducting health impact assessment (HIA) of proposed policies, programs, and projects (for example, transportation, land use, housing, agriculture) at federal, state, tribal, and local levels, including the private sector. The committee will assess the value and potential value of such assessments; the impediments and countervailing factors that have limited the practice of HIA to date; the circumstances and criteria for conducting them; the concepts, tools, and information required; and the types, structure, and content of HIAs. Based on these considerations, the committee will develop a systematic, conceptual framework and approach for improving the assessment of health impacts in the United States.

Appendix D

Glossary

Capacity building: The process by which skills and competence are built for understanding the use for and carrying out a health impact assessment. It may include "policy seminars to sensitise senior managers and advocate change; training courses to build knowledge of method and procedure; dissemination; institutionalisation to enable self-sustaining training in institutions…;[and]case studies and research to build specialist skills."[1]

Community: In the context of this report, the committee uses this term to describe "groups of people who live in the same geographical area; groups of people with a shared history, culture, language; [or] citizens for whom governments are responsible and to whom governments are accountable."[2]

Comprehensive plans: "A legal document that states the goals, principles, policies, and strategies to regulate the growth and development of a particular community… The main characteristics are comprehensiveness, long-range time frame, and holistic territorial coverage. They include elements on land use, economic development, housing, circulation and transportation infrastructures, recreation and open space, community facilities, and community design, among many other possible elements."[3]

---

[1] Birley, M.H. 2001. Annex 3: HIA Guidelines and capacity building. Pp. 39-56 in Health Impact Assessment. WHO/SDE/WSH/01.07. Geneva: World Health Organization [online]. Available: http://hia.anamai.moph.go.th/nwha/pdf/thai62e.pdf [accessed June 8, 2011].

[2] AIDA (Australian Indigenous Doctors' Association). 2010. HIA Connect [online]. Available: http://www.hiaconnect.edu.au/reports/AIDA_HIA.pdf [accessed June 13, 2011].

[3] Hutchinson, E.R., ed. 2010. Pp. 304-305 in Encyclopedia of Urban Studies. Thousand Oaks, CA: Sage Publications. Examples of how comprehensive plans have addressed public health concerns can be found at http://www.planning.org/research/public health/pdf/surveyreport.pdf.

Consultation: "The dynamic process of dialogue between individuals or groups, based upon a genuine exchange of views, and normally with the objective of influencing decisions, policies, or programs of action."[4]

Cost-benefit analysis: A method of considering the advantages and disadvantages of alternative policies or programs by converting all outcomes into monetary values.[5]

Cost-effectiveness analysis: An analysis that compares two or more policies or programs on at least two attributes, for example, costs and benefits. The analysis is done at the margin—that is, to determine the incremental cost effectiveness of one policy or program compared with another, the analyst determines the additional cost required to achieve an additional unit of benefit.[6]

Council on Environmental Quality: An agency in the Executive Office of the President that "coordinates federal environmental efforts and works closely with agencies and other White House offices in the development of environmental policies and initiatives. CEQ was established…by Congress as part of the National Environmental Policy Act of 1969 and additional responsibilities were provided by the Environmental Quality Improvement Act of 1970."[7]

Determinants of health: Many factors contribute to the health of individuals or communities. "Whether people are healthy or not, is determined by their circumstances and environment. To a large extent, factors such as where we live, the state of our environment, genetics, our income and education level, and our relationships with friends and family all have considerable impacts on health, whereas the more commonly considered factors such as access and use of health care services often have less of an impact. The determinants of health include:

---

[4] RTPI (Royal Town Planning Institute). 2005. Guidelines on Effective Community Involvement and Consultation. Royal Town Planning Institute [online]. Available: http://www.rtpi.org.uk/download/385/Guidlelines-on-effective-community-involvement.pdf [accessed June 8, 2011].

[5] Bergus, G.R., S.B. Cantor, M.H. Ebell, T.G. Ganiats, P.P. Glasziou, M.D. Hagen, R.M. Hamm, F.H. Lawler, and J.F. Murray. 1995. A glossary of medical decision-making terms. Prim. Care 22(2):385-393.

[6] Bergus, G.R., S.B. Cantor, M.H. Ebell, T.G. Ganiats, P.P. Glasziou, M.D. Hagen, R.M. Hamm, F.H. Lawler, and J.F. Murray. 1995. A glossary of medical decision-making terms. Prim. Care 22(2):385-393.

[7] CEQ (Council on Environmental Quality). 2010. The Council on Environmental Quality – About. Council on Environmental Quality [online]. Available: http://www.whitehouse.gov/administration/eop/ceq/about [accessed Nov. 22, 2010].

the social and economic environment, the physical environment, and the person's individual characteristics and behaviours."[8]

Environmental assessment (EA): In the context of the National Environmental Policy Act, an environmental assessment is a public document that briefly discusses a proposed action and alternatives to it, including the need for the action and the direct, indirect, and cumulative ecologic, cultural, historical, social, or health impacts of the proposed action and the alternatives. It may be the basis for determining whether the proponent agency has a responsibility for preparing a more comprehensive environmental impact statement or whether it can execute a finding of "no significant impact." It also aids in an agency's compliance with the statute when an environmental impact statement is not necessary.[9]

Environmental impact assessment (EIA): "The process of identifying, predicting, evaluating and mitigating the biophysical, social, and other relevant effects of development proposals prior to major decisions being taken and commitments made."[10] It is a process mandated by law in countries around the world, including the United States, and is also used by multilateral development banks.

Environmental impact statement (EIS): The "detailed statement" required by the National Environmental Policy Act for proposed major federal actions "significantly affecting the quality of the human environment."[11] It is prepared prior to a federal agency making a decision on the proposed action and must include an analysis of the effects of the proposed action and reasonable alternatives to it.[12]

Environmental, social, and health impact assessment (ESHIA): An integrated process by which the impacts of a project on the environment, society, and the health of individuals and the surrounding community are assessed. These assessments are currently carried out more often in the oil, gas, and mining industries.[13, 14, 15]

---

[8]WHO (World Health Organization). 2011. The Determinants of Health. Health Impact Assessment. World Health Organization [online]. Available: http://www.who.int/hia/evidence/doh/en/ [accessed Feb. 10, 2011].

[9]40 C.F.R. §1508.9.

[10]International Association for Impact Assessment. 1999. Principles of Environmental Impact Assessment Best Practice. International Association for Impact Assessment [online]. Available: http://www.iaia.org/publicdocuments/special-publications/Principles%20of%20IA_web.pdf [accessed Nov. 22, 2010].

[11]74 Fed. Reg. 63765 [2009].

[12]42 U.S.C. Section 4332 (1969).

[13]IPIECA/OGP (International Petroleum Industry Environmental Conservation Association and International Association of Oil and Gas Producers). 2007. Health Perform-

European Union: In 2010, the European Union (EU) had 27 member states and four applicants for membership. The EU "is not a federation like the United States. Nor is it simply an organisation for the co-operation between governments, like the United Nations. The countries that make up the EU (its 'Member States') remain independent sovereign nations but they pool their sovereignty...and delegate some of their decision-making powers to shared institutions."[16] "The EU's decision-making process in general and the co-decision procedure in particular involve three main institutions: the European Parliament (EP), which represents the EU's citizens and is directly elected by them; the Council of the European Union, which represents the individual member states; [and] the European Commission, which seeks to uphold interests of the Union as a whole."[17] The commission proposes new laws, which are debated and then adopted by the European Parliament and the council of the EU. The commission and the member states then implement the laws, and the commission ensures that the laws are properly carried out.[18]

Framework: A set of basic elements of a process for evaluating scientific and technical information; in the context of HIA, this process is conducted to understand the potential adverse and beneficial effects of proposed policies, plans, programs, and projects on health.

Health: "A state of complete physical, mental and social well-being and not merely the absence of disease or infirmity."[19]

---

ance Indicators: A Guide for the Oil and Gas Industry. OGP Report No. 393. International Petroleum Industry Environmental Conservation Association, and International Association of Oil and Gas Producers [online]. Available: http://www.ipieca.org/system/files/publications/HPI.pdf [accessed June 2, 2011].

[14]ICMM (International Council on Mining and Metals). 2010. Good Practice Guidance on Health Impact Assessment. London, UK: International Council on Mining and Metals [online]. Available: http://www.icmm.com/page/35457/good-practice-guidance-on-health-impact-assessment [accessed May 16, 2011].

[15]Chevron. 2011. Stakeholder Engagement. Growing Successful Partnerships. Highlights [online]. Available: http://www.chevron.com/globalissues/corporateresponsibility/2007/stakeholderengagement/#b2 [accessed Feb. 10, 2011].

[16]EC (The European Commission). 2010. How the EU Works. The European Commission [online]. Available: http://ec.europa.eu/ireland/about_the_eu/how_the_eu_works/index_en.htm [accessed February 11, 2011].

[17]EU (European Union). 2011. EU Institutions and Other Bodies. Europa [online]. Available: http://europa.eu/institutions/index_en.htm [accessed Feb. 11, 2011].

[18]EC (European Commission). 2007. How the European Works: Your Guide to the EU Institutions. European Commission. July 2007 [online]. Available: http://ec.europa.eu/publications/booklets/eu_glance/68/en.doc [accessed Feb. 11, 2011].

[19]WHO (World Health Organization). 2003. WHO Definition of Health. World Health Organization [online]. Available: http://www.who.int/about/definition/en/print.html [accessed Nov. 22, 2010].

Appendix D

Health disparities: "Systematic, plausibly avoidable health differences adversely affecting socially disadvantaged groups."[20]

Health effect, health impact: In this report, these two terms are used interchangeably and defined as any change in the health of a population or subpopulation or any change in the physical, natural, or cultural environment that has a bearing on public health.

Health impact assessment: The most commonly cited definition of health impact assessment (HIA) is in the Gothenburg consensus paper:

> A combination of procedures, methods and tools by which a policy, program or project may be judged as to its potential effects on the health of a population, and the distribution of those effects within the population.[21]

Other definitions have arisen over the decades, and several examples are provided in Chapter 1, Table 1-1. As discussed in Chapter 3, the committee has chosen to adapt the International Association of Impact Assessment definition[22] and define HIA as follows:

> HIA is a systematic process that uses an array of data sources and analytic methods and considers input from stakeholders to determine the potential effects of a proposed policy, plan, program, or project on the health of a population and the distribution of those effects within the population. HIA provides recommendations on monitoring and managing those effects.

The committee has selected a six-step framework as the clearest way to organize and describe the critical elements of an HIA (see Chapter 3).

> *Screening* determines whether a proposal is likely to have health effects and whether the HIA will provide information useful to the stakeholders and decision-makers.

---

[20] Braveman, P.A., S. Kumanyika, J. Fielding, T. LaVeist, L.N. Borrell, R. Manderscheid, and A. Troutman. 2011. Health disparities and health equity: The issue is justice. American Journal of Public Health [online]. Available: http://ajph.aphapublications.org/cgi/reprint/AJPH.2010.300062v1?view=long&pmid=21551385 [accessed July 6, 2011].

[21] WHO (World Health Organization). 1999. P. 4 in Health Impact Assessment: Main Concepts and Suggested Approach. The Gothenburg Consensus Paper. Brussels: European Centre for Health Policy, WHO Regional Office for Europe, Brussels.

[22] Quigley, R., L. den Broeder, P. Furu, A. Bond, B. Cave, and R. Bos. 2006. Health Impact Assessment: International Best Practice Principles. Special Publication Series No. 5. Fargo: International Association for Impact Assessment. September 2006 [online]. Available: http://www.iaia.org/publicdocuments/special-publications/SP5.pdf [accessed May 6, 2011].

Scoping establishes the scope of health effects that will be included in the HIA, the populations affected, the HIA team, sources of data, methods to be used, and alternatives to be considered.

Assessment involves a two-step process that first describes the baseline health status of the affected population and then assesses potential impacts.

Recommendations suggest design alternatives that could be implemented to improve health or actions that could be taken to manage the health effects, if any, that are identified.

Reporting documents and presents the findings and recommendations to stakeholders and decision-makers.

Monitoring and evaluation are variably grouped and described. Monitoring can include monitoring of the adoption and implementation of HIA recommendations or monitoring of changes in health or health determinants. Evaluation can address the process, impact, or outcomes of an HIA.

Health impact assessment (HIA) practitioner: One who conducts HIA as an individual or part of a team.

Health in all policies: "An approach that looks at all public- and private-sector policy making through a health lens, with the objective of promoting and protecting the health of the population by addressing the social and physical environment influences on health."[23]

Human health risk assessment: A process used to incorporate the understanding of the health implications of exposures, often environmental, into the regulatory decision-making process. See the description of "risk assessment" for more information.

Indigenous: "An official definition of 'indigenous' has not been adopted by any UN-system body. Instead the system has developed a modern understanding of this term based on the following: self-identification as indigenous peoples at the individual level and accepted by the community as their member; historical continuity with pre-colonial and/or pre-settler societies; strong link to territories and surrounding natural resources; distinct social, economic or political systems;

---

[23]PHI (Public Health Institute). 2010. PHI statement on Health in all Policies Task Force, March 12, 2010 [online]. Available: http://www.phi.org/news_events/phi_statements.html [accessed Feb. 10, 2011].

distinct language, culture and beliefs; form non-dominant groups of society; [and] resolve to maintain and reproduce their ancestral environments and systems as distinctive peoples and communities."[24]

Land-use planning: Considers a "community's vision for future development; the policies, goals, principles, and standards upon which the development of the community are based; the proposed location, extent, and intensity of future land usage; existing and anticipated future housing needs; the location and types of transportation required; the location of public and private utilities; and the location of educational, recreational, and cultural facilities including libraries, hospitals, and fire and police stations."[25]

Life-cycle assessment (LCA): "A technique to assess the environmental aspects and potential impacts associated with a product, process, or service, by: compiling an inventory of relevant energy and material inputs and environmental releases; evaluating the potential environmental impacts associated with identified inputs and releases; [and] interpreting the results to help you make a more informed decision."[26] "The major stages in [a life-cycle assessment] study are raw material acquisition, materials manufacture, production, use/reuse/maintenance, and waste management."[27]

National Environmental Policy Act: A U.S. federal law that requires federal agencies in the executive branch to "integrate environmental values into their decision-making processes by considering the environmental impacts of their proposed actions and reasonable alternatives to those actions."[28] It establishes U.S. environmental policy and the Council on Environmental Quality.[29]

---

[24] United Nations Permanent Forum on Indigenous Issues. 2006. Indigenous Peoples and Identity. Fact Sheet 1. United Nations Permanent Forum on Indigenous Issues [online]. Available: http://www.un.org/esa/socdev/unpfii/documents/5session_factsheet1.pdf [accessed Jan. 4, 2011].

[25] Breslow, L. 2002. Pp. 677-678 in Encyclopedia of Public Health, Vol. 3. New York: Macmillan.

[26] EPA (U.S. Environmental Protection Agency). 2011. Life-Cycle Assessment (LCA). U.S. Environmental Protection Agency [online]. Available: http://www.epa.gov/nrmrl/lcaccess/ [accessed Feb. 10, 2011].

[27] EPA (U.S. Environmental Protection Agency). 2011. LCA 101. U.S. Environmental Protection Agency [online]. Available: http://www.epa.gov/nrmrl/lcaccess/lca101.html [accessed May 10, 2011].

[28] FedCenter. 2010. NEPA: General Description. FedCenter [online]. Available: http://www.fedcenter.gov/assistance/facilitytour/construction/nepa/ [accessed June 13, 2011].

[29] 42 U.S.C. § 4321 et seq.

Participation: The overarching term that describes "the extent and nature of activities undertaken by those who take part in public or community involvement, [engagement, and consultation.]"[30]

Plan: In the context of this report, a document, often adopted by a government entity, that describes a future course of action for a community to achieve a desired vision or goal. A plan typically describes the vision and goals of a community or a problem that must be solved, includes a systematic synthesis of available information to analyze the problem, and identifies future actions that must be taken and future investments that must be made to address the stated problem and achieve the desired vision. Plans are prepared and implemented by all levels of government but are especially common at local government levels. Plans include general or comprehensive plans, land-use plans, economic-development plans, and transportation plans. Plans that are commonly subjected to health impact assessment include plans for land use, infrastructure, and natural-resource management.

Policy: Generally, "an agreement or consensus on a range of issues, goals and objectives which need to be addressed....For example, 'Saving Lives: Our Healthier Nation' can be seen as a national health policy aimed at improving the health of the population of England, reducing health inequalities and setting objectives and targets which can be used to monitor progress towards the policy's overall goal or aims."[31] In the committee's report, the use of the term is extended to refer to anything other than land-use plans or development and infrastructure projects. In this context, policy includes formal and informal social rules, including legislation, regulation, budgets, guidelines, and practices.

Program: "Usually refers to a group of activities which are designed to be implemented in order to reach policy objectives.... For example, many Single Regeneration Budget programmes and New Deal for Communities initiatives have a range of themes within their programmes—often including health, community safety (crime), education, employment and housing—and within these themes

---

[30]RTPI (Royal Town Planning Institute). 2005. Guidelines on Effective Community Involvement and Consultation. Royal Town Planning Institute [online]. Available: http://www.rtpi.org.uk/download/385/Guidlelines-on-effective-community-involvement.pdf [accessed June 8, 2011].

[31]WHO (World Health Organization). 2011. Health Impact Assessment (HIA) Glossary of Terms Used [online]. Available: http://www.who.int/hia/about/glos/en/index1.html [accessed Feb. 11, 2011].

Appendix D *193*

are a number of specific projects which, together, make up the overall programme."[32]

Project: "Usually a discrete piece of work addressing a single population group or health determinant, usually with a pre-set time limit."[33] "Usually (but not always), the term refers to 'bricks and mortar' projects involving construction of a discrete structure or group of structures, such as a power plant, highway, or housing development."[34]

Public (or community) engagement: Action taken to begin to "establish effective relationships with individuals or groups so that more specific interactions can then take place."[35]

Public health: The Institute of Medicine has defined public health as "what we, as a society, do collectively to assure the conditions in which people can be healthy."[36] However, the term used in the present report refers more generally to the health of the public. This use is synonymous with the emerging term *population health*.[37] Implicit in both terms is the notion that health is affected by a wide array of factors that range from the societal to the biologic.

Public (or community) involvement: "Effective interactions between planners, decision-makers, individual and representative stakeholders to identify issues and to exchange views on a continuous basis."[38]

---

[32]WHO (World Health Organization). 2011. Health Impact Assessment (HIA) Glossary of Terms Used [online]. Available: http://www.who.int/hia/about/glos/en/index1.html [accessed Feb. 11, 2011].

[33]WHO (World Health Organization). 2011. Health Impact Assessment (HIA) Glossary of Terms Used [online]. Available: http://www.who.int/hia/about/glos/en/index1.html [accessed Feb. 11, 2011].

[34]UCLA HI-CLIC (University of California, Los Angeles-Health Impact Assessment Clearinghouse Learning and Information Center). 2011. Glossary [online]. Available: http://www.hiaguide.org/glossary [accessed Feb. 11, 2011].

[35]RTPI (Royal Town Planning Institute). 2005. Guidelines on Effective Community Involvement and Consultation. Royal Town Planning Institute [online]. Available: http://www.rtpi.org.uk/download/385/Guidlelines-on-effective-community-involvement.pdf [accessed June 8, 2011].

[36]IOM (Institute of Medicine). 1988. The Future of Public Health. Washington, DC: National Academy Press.

[37]Kindig, D.A. 2007. Understanding population health terminology. Milbank. Q 85 (1):139-161.

[38]RTPI (Royal Town Planning Institute). 2005. Guidelines on Effective Community Involvement and Consultation. Royal Town Planning Institute [online]. Available: http://www.rtpi.org.uk/download/385/Guidlelines-on-effective-community-involvement.pdf [accessed June 8, 2011].

Risk assessment: Traditionally, risk assessment is defined as "the characterization of the potential adverse health effects of human exposures to environmental hazards." Risk assessment can be divided into four major steps: hazard identification ("the process of determining whether exposure to an agent can cause an increase in the incidence of a health condition"), dose-response assessment ("the process of characterizing the relation between the dose of an agent administered or received and the incidence of an adverse health effect in exposed populations and estimating the incidence of effect as a function of human exposure to the agent"), exposure assessment ("the process of measuring or estimating the intensity, frequency, and duration of human exposures to an agent"), and risk characterization ("the process of estimating the incidence of a health effect under the various conditions of human exposure described in exposure assessment").[39]

Stakeholder: Any individual or group that will be affected by the outcome of a decision. Stakeholders may include the affected community or specific interest groups, individuals, or organizations that have an economic stake in the outcome and the proponents of a project.[40]

State environmental policy act: Legislation that "provides a way to identify possible environmental impacts that may result from governmental decisions [at the state-level]. These decisions may be related to issuing permits for private projects, constructing public facilities, or adopting regulations, policies or plans."[41] Several states have state environmental policy acts, including California, Connecticut, North Carolina, Washington, and Wisconsin.

Strategic environmental assessment (SEA): "A systematic and anticipatory process, undertaken to analyze the environmental effects of proposed government plans, programmes and other strategies, and to integrate the findings into decision-making. It involves the public and environmental and health authorities, giving them a say in government planning: the responsible authority has to arrange for informing the public and consulting the public concerned, and the decision-maker has to take due account of comments received from the public and from the environmental and health authorities. Such assessments are most commonly carried out for land-use planning at various levels of government, but

---

[39] NRC (National Research Council). 1983. Risk Assessment in the Federal Government: Managing the Process. Washington, DC: National Academy Press.

[40] Mindell, J., E. Ison, and M. Joffe. 2003. A glossary for health impact assessment. J. Epidemiol. Community Health. 57(9):674-651.

[41] Washington State Department of Ecology. 2002. Washington State Environmental Policy Act. Publication No. 02-06-013. FOCUS Sheet May 2002 [online]. Available: http://www.ecy.wa.gov/pubs/0206013.pdf [accessed Mar. 22, 2011].

Appendix D

are also applied to other sectoral plans, such as for energy, water, waste, transport, agriculture and industry."[42]

Tribal environmental policy act: A model act that would establish an environmental impact assessment for actions proposed by tribal governments in the United States.[43]

Zoning ordinance (or bylaws): "Legislative regulations by which a municipal government seeks to control the use of buildings and land within the municipality. It has become, in the United States, a widespread method of controlling urban and suburban construction and removing congestion and other defects of existing plans."[44]

---

[42]UNECE (United Nations Economic Commission for Europe). 2010. New International Treaty to Better Integrate Environmental and Health Concerns into Political Decision-Making. United Nations Economic Commission for Europe. July 6, 2010[online]. Available: http://www.unece.org/press/pr2010/10env_p22e.htm [accessed Jan. 3, 2011].

[43]The Tulalip Tribes of Washington. 2000. Participating in the National Environmental Policy Act: Developing a Tribal Environmental Policy Act. A Comprehensive Guide for American Indian and Alaska Native Communities. The Tulalip Tribes of Washington [online]. Available: http://www.tulalip.nsn.us/pdf.docs/Tribal_EA_Handbook.pdf [accessed Nov. 22, 2010].

[44]Columbia University. 2007. The Columbia Electronic Encyclopedia, 6th Ed. New York: Columbia University Press.

Appendix E

Summary of Health Impact Assessment Guides

Tables E-1 and E-2 provide a summary of health impact assessment (HIA) guides for each stage of the HIA process. Specifically, Table E-1 examines how HIA guides conceptualize the stages of an HIA. It does not review emerging approaches—such as practice standards (Bhatia et al. 2009, 2010)—or review criteria (Fredsgaard et al. 2009). Table E-2 provides an overview of HIA guides for policies and plans.

TABLE E-1 A Review of Health Impact Assessment Guides[a]

	Screening	Scoping	Assessment	Reporting	Recommendations	Monitoring and Evaluation
Development Lending (Pfeiffer and Dora, unpublished material, 2010)[b]	• Screening	• Scoping • Stakeholder engagement	—	• Reviewing HIA report • Reviewing community health action plan	—	• Monitoring community health performance of project
ICMM (International Council on Mining and Metal 2010)	• Screening	• Scoping • Community profiling and baseline studies • Stakeholder and community involvement • Health impact evidence gathering	• Analysis of health impacts • Development of mitigation and enhancement measures	• HIA reporting	—	• Developing health management plan and follow up (monitoring and evaluation)
IFC (International Finance Corporation 2009)	• Screening	• Scoping	• Risk assessment	—	• Health action plan	• Implementation and monitoring • Evaluation and verification of performance and effectiveness
UCLA (Fielding and Cole 2008)	• Screening	• Scoping • Profiling	• Assessment	• Reporting and monitoring	—	—

(Continued)

197

TABLE E-1 Continued

	Screening	Scoping	Assessment	Reporting	Recommendations	Monitoring and Evaluation
MWIA (Coggins et al. 2008)	• Screening • Evidence based assessment	• Scoping	• Appraisal process • Community profiling • Stakeholder and key informant • Research such as literature search	• Identification of potential beneficial or adverse impacts	• Identification of recommendations and writing of report	• Identification of indicators for monitoring impacts of proposal on mental well-being and implementation of recommendations
CHETRE (Harris et al. 2007)	• Screening	• Scoping • Identification	• Assessment	—	• Decision-making and recommendations	• Evaluation and follow-up
Greenspace (Greenspace Scotland 2008)	• Screening • Set up a team to do HIA	• Scoping • Local profile • Involve stakeholders	• Identification and assessment of impacts		• Make recommendations	• Monitor impacts
IAIA (Quigley et al. 2006)	• Screening	• Scoping	• Full-scale HIA • Public engagement and dialogue	• Appraisal of HIA report	• Establishment of framework for intersectoral action • Negotiation of resource allocations for health safeguard measures	• Monitoring of compliance
IPIECA (IPIECA/OGP 2005)	• Screening	• Scoping	• Risk assessment; impact assessment	• Decision-making; establishing priorities; reporting	—	• Implementation and monitoring • Evaluation

EFHIA (Mahoney et al. 2004)	• Screening	• Scoping • Impact identification	• Assessment of impacts	—	• Recommendations	• Evaluation and monitoring
Merseyside (Scott-Samuel et al. 2001)	• Screening • Establish steering group • Agree on terms of reference for assessment • Select assessor	—	• Conduct assessment[c]	• Appraise assessment	• Negotiate favored options	• Implement and monitor • Evaluate and document
EHIA (Fehr 1999)	• Project analysis • Regional analysis	• Population analysis • Background situation	• Prognosis of future pollution • Summary assessment of impacts	—	• Recommendations • Communication	• Evaluation

[a]This table examines how HIA guides conceptualize the stages of an HIA. It does not review emerging approaches, such as practice standards (Bhatia et al. 2009, 2010), or review criteria (Fredsgaard et al. 2009).
[b]This is not strictly a guide, and there is no assessment stage. The document assists lenders in following and reviewing health assessments.
[c]This stage includes seven steps and covers scoping and assessment.
Abbreviations: CHETRE, Centre for Health Equity Training, Research and Evaluation; EFHIA, equity-focused health impact assessment; EHIA, environmental health impact assessment; IAIA, International Association for Impact Assessment; ICMM, International Council on Mining and Metals; IFC, International Finance Corporation; IPIECA, International Petroleum Industry Environmental Conservation Association; MWIA, mental well-being impact assessment; and UCLA, University of California, Los Angeles.

TABLE E-2 Health Impact Assessment Guides for Policies or Plans

	SEA	EPHIA
	(EP/Council 2001; ODPM 2005)	(Abrahams et al. 2004)
Screening	Screening—2.5 under Article 2(a), the plans and programs subject to the directive are those which are: • Subject to preparation or adoption by an authority at national, regional, or local level or prepared by an authority for adoption through a legislative procedure by Parliament or government. • Required by legislative, regulatory, or administrative provisions.	Screening
Scoping	Stage A—Setting the context and objectives, establishing the baseline, deciding on scope A1: Identifying relevant plans, programs, and environmental protection objectives. A2: Collecting baseline data. A3: Identifying environmental problems. A4: Developing SEA objectives, indicators, and targets. A5: Consulting on the scope of SEA.	Scoping
Assessment	Stage B—Alternatives and assessment B1: Testing the plan or program objectives against the SEA objectives. B2: Developing strategic alternatives. B3: Predicting the effects of the draft plan or program, including alternatives. B4: Evaluating the effects of the draft plan or program, including alternatives. B5: Considering ways to mitigate adverse effects.	• Conduct assessment • Policy analysis • Qualitative and quantitative data collection • Impact analysis • Setting priorities among impacts • Recommendations developed • Profiling

	B6: Proposing measures to monitor the environmental effects of plan or program implementation.	• Process evaluation
Reporting	Stage C—Preparing the environmental report	Report on health impacts and policy options
Recommendations	Stage D—Consultation and decision-making Responsible authorities will • Consult on the draft plan or programme and the environmental report. • Assess significant changes.	
Monitoring and evaluation	Stage E—Monitoring implementation of the plan or program • Developing aims of and methods for monitoring.	• Monitoring • Impact and outcome evaluation

Abbreviations: EPHIA, European policy health impact assessment; and SEA, strategic environmental assessment.

REFERENCES

Abrahams, D., A. Pennington, A. Scott-Samuel, C. Doyle, O. Metcalfe, L. den Broeder, F. Haigh, O. Mekel, and R. Fehr. 2004. European Policy Health Impact Assessment: A Guide, University of Liverpool, England; RIVM, Netherlands; Institute of Public Health, Ireland; loegd, Institute of Public Health, NRW Bielefeld, Germany. Prepared for the Health and Consumer Protection Directorate General, European Commission. May 2004 [online]. Available: http://ec.europa.eu/health/ph_projects/2001/monitoring/fp_monitoring_2001_a6_frep_11_en.pdf [accessed May 16, 2011].

Bhatia, R., L. Farhang, M. Gaydos, K. Gilhuly, B. Harris-Roxas, J. Heller, M. Lee, J. McLaughlin, M. Orenstein, E. Seto, L. St. Pierre, A.L. Tamburrini, A. Wernham, and M. Wier. 2009. Practice Standards for Health Impact Assessment (HIA), Version 1. North American HIA Practice Standards Working Group, Oakland, CA. April 2009 [online]. Available: http://www.habitatcorp.com/whats_new/HIA_Practice_Standards_040709_V1.pdf [accessed May 17, 2011].

Bhatia, R., J. Branscomb, L. Farhang, M. Lee, M. Orenstein, and M. Richardson. 2010. Minimum Elements and Practice Standards for Health Impact Assessment (HIA), Version 2. North American HIA Practice Standards Working Group, Oakland, CA. November 2010 [online]. Available: http://www.sfphes.org/HIA_Tools/HIA_Practice_Standards.pdf [accessed May 23, 2011].

Coggins, T., A. Cooke, L. Friedli, J. Nicholls, A. Scott-Samuel, and J. Stansfield. 2008. Mental Well-Being Impact Assessment: A Toolkit, "A Living and Working Document". Care Services Improvement Partnership, North West Development Centre [online]. Available: http://www.liv.ac.uk/ihia/IMPACT%20Reports/mwia-toolit1.pdf [accessed May 16, 2011].

EP/Council (European Parliament and Council of the European Union). 2001. Directive 2001/42/EC of the European Parliament and of the Council of 27 June 2001 on the assessment of the effects of certain plans and programmes on the environment. O.J. Eur. Comm. L 197:30-37.

Fehr, R. 1999. Environmental health impact assessment: Evaluation of a 10 step model. Epidemiology 10(5):618-625.

Fielding, J., and B.L. Cole. 2008. UCLA Training Manual. Health Impact Assessment Clearinghouse Learning and Information Center [online]. Available: http://www.ph.ucla.edu/hs/health-impact/training.htm#uclatraining [accessed May 19, 2011].

Fredsgaard, M.W., B. Cave, and A. Bond. 2009. A Review Package for Health Impact Assessment Reports of Development Projects. Leeds, UK: Ben Cave Associates Ltd [online]. Available: http://www.bcahealth.co.uk/pdf/hia_review_package.pdf.

Greenspace Scotland. 2008. Health Impact Assessment of Greenspace: A Guide. Health Scotland, Greenspace Scotland, Scottish Natural Heritage and Institute of Occupational Medicine. Stirling: Greenspace Scotland. June 2008 [online]. Available http://www.greenspacescotland.org.uk/upload/File/Greenspace%20HIA.pdf [accessed May 17, 2011].

Harris, P., B. Harris-Roxas, E. Harris, and L. Kemp. 2007. Health Impact Assessment: A Practical Guide. Sidney, Australia: Centre for Health Equity Training, Research and Evaluation, the University of New South Wales. August 2007 [online]. Available: http://www.hiaconnect.edu.au/files/Health_Impact_Assessment_A_Practical_Guide.pdf [accessed May 9, 2011].

ICMM (International Council on Mining and Metals). 2010. Good Practice Guidance on Health Impact Assessment. London, UK: International Council on Mining and Metals [online]. Available: http://www.icmm.com/page/35457/good-practice-guidance-on-health-impact-assessment [accessed May 16, 2011].

IFC (International Finance Corporation). 2009. Introduction to Health Impact Assessment. Washington, DC: World Bank [online]. Available: http://www.ifc.org/ifcext/sustainability.nsf/AttachmentsByTitle/p_HealthImpactAssessment/$FILE/HealthImpact.pdf [accessed May 5, 2011].

IPIECA/OGP (International Petroleum Industry Environmental Conservation Association and International Association of Oil and Gas Producers). 2005. A Guide to Health Impact Assessments in the Oil and Gas Industry. International Petroleum Industry Environmental Conservation Association, and International Association of Oil and Gas Producers [online]. Available: http://www.hiaconnect.edu.au/files/HIA_in_OG.pdf [accessed May 17, 2011].

Mahoney, M., S. Simpson, E. Harris, R. Aldrich, and J. Stewart-Williams. 2004. Equity Focused Health Impact Assessment Framework. The Australasian Collaboration for Health Equity Impact Assessment (ACHEIA). August 2004 [online]. Available: http://www.hiaconnect.edu.au/files/EFHIA_Framework.pdf [accessed May 17, 2011].

ODPM (Office of the Deputy Prime Minister). 2005. A Practical Guide to the Strategic Environmental Assessment Directive. Department for Communities and Local Governments [online]. Available: http://www.communities.gov.uk/documents/planningandbuilding/pdf/practicalguidesea.pdf [accessed June 10, 2011].

Quigley, R., L. den Broeder, P. Furu, A. Bond, B. Cave, and R. Bos. 2006. Health Impact Assessment: International Best Practice Principles. Special Publication Series No. 5. Fargo: International Association for Impact Assessment. September 2006 [online]. Available: http://www.iaia.org/publicdocuments/special-publications/SP5.pdf [accessed May 6, 2011].

Scott-Samuel, A., M. Birley, and K. Ardern. 2001. The Merseyside Guidelines for Health Impact Assessment, 2nd Ed. Liverpool: International Health Impact Assessment Consortium. May 2001 [online]. Available: http://www.hiaconnect.edu.au/files/Merseyside_Guidelines.pdf [accessed May 18, 2011].

Appendix F

Analysis of Health Effects under the National Environmental Policy Act

In Chapter 4, the committee noted that the analysis of health effects under the National Environmental Policy Act (NEPA) has been limited. To date, neither the Council on Environmental Quality (CEQ) nor federal agencies that comply with NEPA have produced guidance on the analysis of health effects. As discussed in Chapter 4, the lack of guidance on analyzing public-health effects does not diminish the legal requirement to consider health in an environmental impact statement (EIS). Agencies complying with NEPA, however, often lack public-health expertise, and the lack of guidance may be a disincentive to a more robust, systematic approach to health. Although there is no formal guidance, existing regulations and relevant guidance provide a foundation for improving the analysis of health effects in an EIS. To assist the agencies in conducting a more robust, systematic analysis of health impacts, this appendix addresses the following issues:

- Determining when to conduct a systematic analysis of health effects in an EIS or environmental assessment.
- Determining the appropriate scope of health problems to include in the analysis.
- Determining what populations or communities are affected and describing baseline conditions in them.
- Analysis of health effects in a manner that is scientifically and legally defensible according to the requirements of NEPA.
- Mitigation of identified effects on public health.
- Responsibility and authority for public-health analysis under NEPA.

DETERMINING WHEN TO CONDUCT AN ANALYSIS OF HEALTH EFFECTS

Health effects should be considered in complying with NEPA (40 CFR 1508.8). However, the CEQ also instructs agencies to "identify and eliminate from detailed study the issues which are not significant or which have been covered by prior environmental review" (40 CFR § 1501.7(a)3). Agencies are thus obliged to consider health effects only when there is reason to conclude that they may be significant. Questions that agencies may wish to answer in determining significance include the following:

- Were scoping comments on health submitted?
- Are health concerns a major point of controversy (even if the concerns that have been raised are not likely to be supported by the analysis)?
- Are there other significant impacts likely that are known to affect health? The effects of federal-agency actions subject to NEPA that may impact health include emissions of hazardous substances; changes in community demographics; involuntary displacement of residents or businesses; changes in industry actions or practices, employment, government revenues, or land-use patterns; changes in modes or safety of transportation; reductions in access to natural resources; and changes in food and agricultural resources.

Although environmental-justice guidance is intended to assist agencies in addressing the potential for disparate effects on low-income and minority-group communities, some of the principles also have relevance to health effects in the general population. The CEQ suggests that agencies should "consider enhancing their outreach" to public-health agencies and clinics (CEQ 1997).

DETERMINING THE APPROPRIATE SCOPE OF HEALTH-EFFECT ANALYSIS

CEQ regulations on implementing NEPA contain several statements that can help to guide an agency's approach to scoping for health effects. First, agencies are instructed to consider direct, indirect, and cumulative effects associated with the proposed action and alternatives (40 CFR § 1508.8). Thus, agencies should not arbitrarily limit consideration to health effects that may be the most obvious or direct (such as those related to emissions or discharges) but should systematically consider the potential for direct, indirect, or cumulative health effects. Health determinants that might be considered and analyzed in the scope of an environmental impact assessment under NEPA would be the same as those considered in HIA and would include such factors as the quality and affordability of housing; access to employment and government revenues; the quality and accessibility of parks, schools, and transportation services; neighborhood safety; exposure to environmental hazards; the quality and affordability of food resources; and the extent and strength of social networks. Moreover, agencies should be responsive to concerns raised by stakeholders during scoping, particularly when health concerns are a matter of controversy (40 CFR § 1501.7, 40 CFR § 1508.27(b)(4).

Environmental-justice guidance (EPA 1998) discusses what is relevant to health effects in the general population and states the following:

> The EPA NEPA analyst should develop a full understanding of baseline demographic, socioeconomic, and environmental conditions so that a comprehensive assessment of the types of impacts that may be imposed upon all human and natural resources...can be conducted and an understanding of how these impacts may translate into human health concerns can be developed.

NEPA and CEQ regulations do not identify any category of health effect that is exempt from consideration under NEPA. Agencies are instructed to include all effects that may be significant, whether direct, indirect, or cumulative. CEQ regulations (40 CFR § 1501.7(a)(3)) do, however, require that agencies do the following:

> Identify and eliminate from detailed study the issues which are not significant or which have been covered by prior environmental review (§1506.3), narrowing the discussion of these issues in the statement to a brief presentation of why they will not have a significant effect on the human environment or providing a reference to their coverage elsewhere.

In practice, a systematic approach to identifying health effects should help agencies to ensure that potentially significant health effects are included.

DETERMINING THE AFFECTED POPULATIONS OR COMMUNITIES AND DESCRIBING THE BASELINE

The description of the affected environment in the regulations indicates the baseline with which impacts of the alternatives can be compared. For public health, the comparison should include a concise discussion of the health status and health determinants in the affected community. CEQ regulations clearly indicate that the EIS should focus on describing aspects of the affected environment that are necessary for developing an understanding of the effects of the alternatives (40 CFR § 1502.15). For public health, therefore, the goal is not a comprehensive assessment of all health issues, but only the ones that are relevant to the health impacts identified.

Public-health data and statistics for describing the public-health environment will be drawn from a variety of sources. Federal, tribal, state, and local health departments maintain databases and surveillance on various health conditions; local hospitals and clinics may also have relevant data. There may be restrictions on accessing or publishing some statistics because health data are subject to legal requirements intended to protect privacy. Consultation with the appropriate health officials is a way for agencies to identify and access appropriate data. Establishing cooperating agency relationships with the relevant health agencies may also be desirable (40 CFR § 1501.6).

Determining what populations or communities may be affected requires an understanding of the pathways through which impacts may occur. The CEQ notes that the context of the decision is important for determining where significant effects would occur; for example, site-specific actions are more likely to have localized effects (40 CFR §1508.27).

ANALYZING THE HEALTH EFFECTS

As noted above, CEQ regulations require that agencies consider "the direct, indirect, and cumulative effects" of the proposed action and alternative and, as noted in Chapter 4, define health as one of the effects that should be included (40 CFR § 1502.16, 40 CFR § 1508.8). They also note that the analysis may include beneficial effects (40 CFR § 1508.8). Agencies are further directed to consider how "economic or social and natural or physical environmental effects are interrelated" (40 CFR § 1508.14).

The regulations and available guidance do not identify specific methods that must be used in analyzing health effects or other effects more commonly included in an EIS. Instead, NEPA simply requires that agencies "utilize a systematic, interdisciplinary approach which will insure the integrated use of the natural and social sciences and the environmental design arts" (Section 102(2)(A)). Agencies are required to "insure the professional integrity, including scientific integrity, of discussions and analyses in environmental impact statements. They shall identify any methodologies used and shall make explicit reference…to sources relied upon for conclusions in the statement" (40 CFR § 1502.24). Thus, although the regulations on NEPA's implementation do not provide specific guidance on methods that should be used to assess health implications, they establish basic standards and expectations (as for all other effects considered in an EIS) regarding a broad-based, interdisciplinary, scientifically sound approach.

Uncertainty of predictions is a common concern in analyzing health effects, but this challenge is common to many effects considered in an EIS. In many cases, controlled studies of a scenario analogous to the action being assessed do not exist, and the agency must make judgments based on uncertain predictions. CEQ guidance addresses the question of uncertainty and states that "the EIS must…make a good faith effort to explain the effects that are not known but are 'reasonably foreseeable'" and that "the agency has the responsibility to make an informed judgment" and "cannot ignore these uncertain, but probable, effects of its decision" (CEQ 1981).

MITIGATATION OF IDENTIFIED EFFECTS ON PUBLIC HEALTH

Agencies are required to consider mitigation measures as part of the alternatives (40 CFR § 1502.14(f)) or in response to any significant effects identified in the analysis (40 CFR § 1502.16(h)). Some existing regulatory standards (such as those established by the Clean Air Act and Clean Water Act) establish health-based thresholds that trigger actions to minimize exposure to specific pollutants. Many impacts included in an EIS—including some health effects—have no such thresholds or regulatory standards. In some cases, the mitigation measures identified may lie outside the jurisdiction of the lead agency or cooperating agencies. The CEQ (1981) has provided guidance on this situation and states the following:

> All relevant, reasonable mitigation measures that could improve the project are to be identified, even if they are outside the jurisdiction of the lead agency or the cooperating agencies, and thus would not be committed as part of the RODs [Records of Decisions] of these agencies. Sections 1502.16(h), 1505.2(c). This will serve to [46 FR 18032] alert agencies or officials who can implement these extra measures, and will encourage them to do so. Because the EIS is the most comprehensive environmental document, it is an ideal vehicle in which to lay out not only the full range of environmental impacts but also the full spectrum of appropriate mitigation.

Health mitigation measures may be implemented not only through regulations or requirements established by the lead agency but through actions taken by a cooperating agency, another government entity, or local, state, or tribal health department or through voluntary actions taken by a project proponent or another stakeholder.

RESPONSIBILITY AND AUTHORITY FOR PUBLIC-HEALTH ANALYSIS UNDER THE NATIONAL ENVIRONMENAL POLICY ACT

Ultimately, compliance with NEPA requirements is the responsibility of the lead agency. As noted previously, however, agencies are directed specifically to use an interdisciplinary approach (40 CFR § 1502.6). CEQ guidance has emphasized the importance of soliciting cooperating agency participation to fulfill this requirement and ensure a complete, efficient analysis (CEQ 2002).

Finally, CEQ requires that the "disciplines of the preparers shall be appropriate to the scope and issues identified in the scoping process" (40 CFR § 1502.6). Thus, when health effects are to be included, agencies should solicit the participation of public-health experts. Local, state, tribal, and federal health agencies often have adequate public-health knowledge and data but may lack familiarity with NEPA and will require orientation on the procedures and approach.

REFERENCES

CEQ (Council on Environmental Quality). 1981. Forty Most Asked Questions Concerning CEQ's National Environmental Policy Act Recommendations. Memorandum for Federal NEPA Liaisons, Federal, State, and Local Officials and Other Persons Involved in the NEPA Process, from Nicholas C. Yost, General Counsel, Council on Environmental Quality, Washington, DC, March 16, 1981 [online]. Available: http://nepa.energy.gov/nepa_documents/TOOLS/GUIDANCE/Volume1/4-1-40_questions.html [accessed July 12, 2011].

CEQ (Council on Environmental Quality). 1997. Environmental Justice: Guidance Under the National Environmental Policy Act. Council on Environmental Quality, Washington, DC [online]. Available: http://ceq.hss.doe.gov/nepa/regs/ej/justice.pdf [accessed July 12, 2011].

CEQ (Council on Environmental Quality). 2002. Cooperating Agencies in Implementing the Procedural Requirements of the National Environmental Policy Act. Memorandum for the Heads of Federal Agencies, from James Connaughton, Chair, Council on Environmental Quality, Washington, DC. January 30, 2002 [online]. Available: http://ceq.hss.doe.gov/nepa/regs/cooperating/cooperatingagenciesmemorandum.html [accessed July 12, 2011].

EPA (U.S. Environmental Protection Agency). 1998. Final Guidance for Incorporating Environmental Justice Concerns in EPA's NEPA Compliance Analyses. U.S. Environmental Protection Agency. April 1998 [online]. Available: http://www.epa.gov/compliance/ej/resources/policy/ej_guidance_nepa_epa0498.pdf [accessed July 12, 2011].